T0155841

ENCYCLOPEDIA OF MATHEMATICS AND ITS APPLICATIONS
Volume 25

Computation and Automata

ENCYCLOPEDIA OF MATHEMATICS
and Its Applications

GIAN-CARLO ROTA, Editor
Massachusetts Institute of Technology

Editorial Board

For other books in this series see page 283

Computation and Automata

Arto Salomaa

University of Turku
Finland

The right of the
University of Cambridge
to print and sell
all manner of books
was granted by
Henry VIII in 1534.
The University has printed
and published continuously
since 1584.

CAMBRIDGE UNIVERSITY PRESS

Cambridge

London   New York   New Rochelle

Melbourne   Sydney

CAMBRIDGE UNIVERSITY PRESS
Cambridge, New York, Melbourne, Madrid, Cape Town,
Singapore, São Paulo, Delhi, Tokyo, Mexico City

Cambridge University Press
The Edinburgh Building, Cambridge CB2 8RU, UK

Published in the United States of America by Cambridge University Press, New York

www.cambridge.org
Information on this title: www.cambridge.org/9780521177337

First published 1985
First paperback edition 2011

A catalogue record for this publication is available from the British Library

Library of Congress Cataloguing in Publication data
Salomaa, Arto.
Computation and automata.
(Encyclopedia of mathematics and its applications; v. 25)
Bibliography: p.
Includes index.
1. Computable functions. 2. Computational complex-
ity. 3. Sequential machine theory. 1. Title.
11. Series.
QA9.59.S25   1985   511   84–17571

ISBN 978-0-521-30245-6 Hardback
ISBN 978-0-521-17733-7 Paperback

Contents

Editor's Statement

A large body of mathematics consists of facts that can be presented and described much like any other natural phenomenon. These facts, at times explicitly brought out as theorems, at other times concealed within a proof, make up most of the applications of mathematics, and are the most likely to survive change of style and of interest.

This ENCYCLOPEDIA will attempt to present the factual body of all mathematics. Clarity of exposition, accessibility to the non-specialist, and a thorough bibliography are required of each author. Volumes will appear in no particular order, but will be organized into sections, each one comprising a recognizable branch of present-day mathematics. Numbers of volumes and sections will be reconsidered as times and needs change.

It is hoped that this enterprise will make mathematics more widely used where it is needed, and more accessible in fields in which it can be applied but where it has not yet penetrated because of insufficient information.

GIAN-CARLO ROTA

Foreword

The last twenty years have witnessed most vigorous growth in areas of mathematical study connected with computers and computer science. The enormous development of computers and the resulting profound changes in scientific methodology have opened new horizons for the science of mathematics at a speed without parallel during the long history of mathematics.

The following two observations should be kept in mind when reading the present monograph. First, various developments in mathematics have directly initiated the "beginning" of computers and computer science. Second, advances in computer science have induced very vigorous developments in certain branches of mathematics. More specifically, the second of these observations refers to the growing importance of discrete mathematics—and we are now witnessing only the very beginning of the influence of discrete mathematics.

Because of reasons outlined above, mathematics plays a central role in the foundations of computer science. A number of significant research areas can be listed in this connection. It is interesting to notice that these areas also reflect the historical development of computer science.

1. The classical *computability theory* initiated by the work of Gödel, Tarski, Church, Post, Turing, and Kleene occupies a central role. This area is rooted in mathematical logic.

2. In the classical *formal language and automata theory* the central notions are those of an automaton, a grammar, and a language. Apart from

developments in area (1), the work of Chomsky on the foundations of natural languages, as well as the work of Post concerning rewriting systems, should be mentioned here. It is, however, fascinating to observe that the modern theory of formal languages and rewriting systems was initiated by the work of the Norwegian mathematician Axel Thue at the beginning of this century!

3. An area initiated in the sixties is *complexity theory*. The performance of an algorithm is investigated. The central notions are those of a tractable and an intractable problem. This area is gaining in importance because of several reasons, one of them being the advances in area (4).

4. Quite recent developments concerning the security of computer systems have increased the importance of *cryptography* to a great extent. Moreover, the idea of public key cryptography is of specific theoretical interest and has drastically changed our ideas concerning what is doable in communication systems.

Areas (1) through (4) constitute the core of the present monograph. Many other important areas dealing with the mathematical foundations of computer science (e.g., semantics and the theory of correctness of programming languages, the theory of data structures, and the theory of data bases) lie beyond the scope of the present monograph and will, hopefully, be presented in other books in this series.

All the areas listed above comprise a fascinating part of contemporary mathematics that is very dynamic in character, full of challenging problems requiring most interesting and ingenious mathematical techniques.

This monograph provides a very good basis for the understanding of these developments. It presents this fascinating modern area of mathematics in a broad and clear perspective. Because everything is developed essentially from the beginning, even an uninitiated reader can use the monograph as an entry to this area. In spite of this, a glimpse of a number of very recent developments is given.

Grzegorz Rozenberg

Acknowledgments

It is difficult to list all persons who have in some way or other contributed to this book. Parts of the manuscript were used as lecture notes for courses given at the universities of Turku and Waterloo. I want to thank the participants in these courses, in particular, Juha Honkala and Sheng Yu. Tero Harju, Juha Honkala, Werner Kuich, Valtteri Niemi, and Grzegorz Rozenberg have read through at least some parts of the manuscript and given very useful comments. Moreover, I have benefited from discussions with or comments from Karel Culik II, Jozef Gruska, Helmut Jürgensen, Juhani Karhumäki, Matti Linna, Hermann Maurer, Martti Penttonen, Keijo Ruohonen, Adi Shamir, Emo Welzl, and Derick Wood. The difficult task of typing the manuscript was performed in an excellent fashion by Elisa Mikkola. I want to thank the publisher for excellent and timely editorial work with both the typescript and proofs. Last but not least, I want to acknowledge the continuing support of my wife, children, and other members of the family. In particular, discussions with Ilokivi and Turzan were always very encouraging, and the whole book would not have been possible without Ketta and Korak.

Arto Salomaa

CHAPTER 1

Introduction: Models of Computation

The basic question in the theory of computing can be formulated in any of the following ways: What is computable? For which problems can we construct effective mechanical procedures that solve every instance of the problem? Which problems possess algorithms for their solutions?

Fundamental developments in mathematical logic during the 1930s showed the existence of *unsolvable* problems: No algorithm can possibly exist for the solution of the problem. Thus, the existence of such an algorithm is a logical impossibility—its nonexistence has nothing to do with our ignorance. This state of affairs led to the present formulation of the basic question in the theory of computing. Previously, people always tried to construct an algorithm for every precisely formulated problem until (if ever) the correct algorithm was found. The basic question is of definite practical significance: One should not try to construct algorithms for an unsolvable problem. (There are some notorious examples of such attempts in the past.)

A *model of computation* is necessary for establishing unsolvability. If one wants to show that no algorithm for a specific problem exists, one must have a precise definition of an algorithm. The situation is different in establishing solvability: It suffices to exhibit some particular procedure that is effective in the intuitive sense. (We use the terms *algorithm* and *effective procedure* synonymously. There are some obvious requirements every intuitively effective procedure has to satisfy. At the moment we do not try to list such requirements.)

We are now confronted with the necessity of formalizing a notion of a model of computation that is general enough to cover all conceivable computers, as well as our intuitive notion of an algorithm. Some initial observations are in order.

Let us assume that the algorithms we want to formalize compute functions mapping the set of nonnegative integers into the same set. Although this is not important at this point, we could observe that our assumption is no essential restriction of generality. This is due to the fact that other input and output formats can be encoded into nonnegative integers.

After having somehow defined our general model of computation, denoted by MC, we observe that each specific instance of the model possesses a finitary description; that is, it can be described in terms of a formula or finitely many words. By enumerating these descriptions, we obtain an enumeration $MC_1, MC_2, \ldots$ of all specific instances of our general model of computation. In this enumeration, each MC_i represents some particular algorithm for computing a function from nonnegative integers into nonnegative integers. Denote by $MC_i(j)$ the value of the function computed by MC_i for the argument value j.

Define a function $f(x)$ by

$$f(x) = MC_x(x) + 1. \tag{1}$$

Clearly, the following is an algorithm (in the intuitive sense) to compute the function $f(x)$. Given an input x, start the algorithm MC_x with the input x and add one to the output.

However, is there any specific algorithm among our formalized MC-models that would compute the function $f(x)$? The answer is no, and the argument is an indirect one. Assume that MC_t would give rise to such an algorithm, where t is some natural number. Hence, for all x,

$$f(x) = MC_t(x). \tag{2}$$

A contradiction now arises by substituting the value t for the variable x in both (1) and (2).

This contradiction, referred to as the *dilemma of diagonalization,* shows that independently of our model of computation—indeed, we did not specify the MC-model in any way—there will be algorithms not formalized by the model.

There is a simple and natural way to avoid the dilemma of diagonalization. We have assumed so far that the MC_i-algorithms are defined everywhere: For all input j, the algorithm MC_i produces an output. This assumption is unreasonable from many points of view, one of which is computer programming; we cannot be sure that every program produces an output for every input. Therefore, we should allow also the possibility that some of the

MC_i-algorithms enter an infinite loop for some inputs j and, consequently, do not produce any output for such a j. Moreover, the set of such values j is not known a priori.

Thus, some algorithms in the list

$$MC_1, MC_2, \ldots \tag{3}$$

produce an output only for some of the possible inputs; that is, the corresponding functions are not defined for all nonnegative integers. The dilemma of diagonalization does not arise after the inclusion of such *partial* functions among the functions computed by the algorithms of (3). Indeed, the argument presented above does not lead to a contradiction because $MC_i(t)$ is not necessarily defined.

The general model of computation, now referred to as a *Turing machine,* was introduced quite a long time before the advent of electronic computers. Turing machines constitute by far the most widely used general model of computation. Other general models discussed later in this book are *Markov algorithms, Post systems, grammars,* and *L systems.* Each of these models leads to a list such as (3), where partial functions are also included. All models are also equivalent in the sense that they define the same set of solvable problems or computable functions. This is understood in a sense made precise later; also, the input and output formats are taken into account. For instance, grammars naturally define languages and, consequently, an input-output format associated to computing function values is rather unsuitable for grammars.

We have considered only the general question of characterizing the class of solvable problems. This question was referred to as basic in the theory of computing. It led to a discussion of general models of computation.

More specific questions in the theory of computing deal with the *complexity* of solvable problems. Is a problem P_1 more difficult that P_2 in the sense that every algorithm for P_1 is more complex (for instance, in terms of time or memory space needed) than a reasonable algorithm for P_2? What is a reasonable classification of problems in terms of complexity? Which problems are so complex that they can be classified as *intractable* in the sense that all conceivable computers require an unmanageable amount of time for solving the problem?

Undoubtedly, such questions are of crucial importance from the point of view of practical computing. A problem is not yet settled if it is known to be solvable or computable and remains intractable at the same time. As a typical example, many recent results in cryptography are based on the assumption that the factorization of the product of two large primes is impossible in practice. More specifically, if we know a large number n consisting of, for example, 200 digits and if we also know that n is the product of two large primes, it is still impossible for us to find the two primes. This assumption is reasonable because the problem described is intractable, at least in view of the factoring algorithms known at present. Of course, from a merely theoretical point of view where complexity is not considered, such a factoring algorithm can be trivially constructed.

Such specific questions lead to more specific models of computing. The latter are obtained either by imposing certain restrictions on Turing machines or else by some direct construction. Also, such specific models will be discussed in the sequel. Of particular importance is the *finite automaton*. It is a model of a strictly finitary computing device: The automaton is not capable of increasing any of its resources during the computation.

It is clear that no model of computation is suitable for all situations; modifications and even entirely new models are needed to match new developments. Theoretical computer science by now has a history long enough to justify a discussion about good and bad models. The theory is mature enough to produce a great variety of different models of computation and prove some interesting properties concerning them. Good models should be general enough so that they are not too closely linked with any particular situation or problem in computing—they should be able to lead the way. On the other hand, they should not be too abstract. Restrictions on a good model should converge, step by step, to some area of real practical significance. A typical example is some restrictions of abstract grammars especially suitable for considerations concerning parsing. The resulting aspects of parsing are essential in compiler construction.

To summarize: A good model represents a well-balanced abstraction of a real practical situation—not too far from and not too close to the real thing.

Formal languages constitute a descriptive tool for models of computation, both in regard to the input-output format and the mode of operation. Formal language theory is by its very essence an interdisciplinary area of science; the need for a formal grammatical description arises in various scientific disciplines, ranging from linguistics to biology. Therefore, appropriate aspects of formal language theory will be of crucial importance in this book.

CHAPTER 2

Rudiments of Language Theory

2.1. LANGUAGES AND REWRITING SYSTEMS

Both natural and programming languages can be viewed as sets of sentences—that is, finite strings of elements of some basic vocabulary. The notion of a language introduced in this section is very general. It certainly includes both natural and programming languages and also all kinds of nonsense languages one might think of. Traditionally, formal language theory is concerned with the syntactic specification of a language rather than with any semantic issues. A syntactic specification of a language with finitely many sentences can be given, at least in principle, by listing the sentences. This is not possible for languages with infinitely many sentences. The main task of formal language theory is the study of finitary specifications of infinite languages.

The basic theory of computation, as well as of its various branches, such as cryptography, is inseparably connected with language theory. The input and output sets of a computational device can be viewed as languages, and—more profoundly—models of computation can be identified with classes of language specifications, in a sense to be made more precise. Thus, for instance, Turing machines can be identified with phrase-structure grammars and finite automata with regular grammars.

We begin by introducing some notions and terminology fundamental to all our discussions.

An **alphabet** is a finite, nonempty set. The elements of an alphabet, which we might call Σ, are referred to as **letters**, or **symbols**. A **word** over an alphabet Σ is a finite string consisting of zero or more letters of Σ, in which the same letter may occur several times. The string consisting of zero letters is called the **empty word**, written λ. For instance, λ, 0, 10, 1011, and 00000 are words over the alphabet $\Sigma = \{0, 1\}$. The set of all words (resp. all nonempty words) over an alphabet Σ is denoted by Σ^* (resp. Σ^+). The sets Σ^* and Σ^+ are infinite for any Σ. Algebraically speaking, Σ^* and Σ^+ are the free monoid (with the identity λ) and the free semigroup generated by Σ.

The reader should keep in mind that the basic set Σ, its elements, and strings of its elements could equally well be called a *vocabulary, words,* and *sentences,* respectively. This would reflect an approach with applications mainly in the area of natural languages. In this book, we use the standard mathematical terminology introduced above.

For words w_1 and w_2, the juxtaposition $w_1 w_2$ is called the **catenation** (or concatenation) of w_1 and w_2. The empty word is an identity with respect to catenation: $w\lambda = \lambda w = w$ holds for all words w. Because catenation is associative, the notation w^i, where i is a positive integer, is used in the customary sense. By definition, w^0 is the empty word, λ.

The **length** of a word w, denoted by $|w|$, is the number of letters in w when each letter is counted as many times as it occurs. Again by definition, $|\lambda| = 0$. The length function possesses some of the formal properties of logarithm:

$$|w_1 w_2| = |w_1| + |w_2|, \qquad |w^i| = i|w|$$

for all words w and integers $i \geq 0$.

A word w is a **subword** (or a *factor*) of a word u if there are words x and y such that $u = xwy$. Furthermore, if $x = \lambda$ (resp. $y = \lambda$), then w is called an **initial** subword, or a **prefix**, of u (resp. a **final** subword or a **suffix** of u).

Subsets of Σ^* are referred to as **formal languages**—or, briefly, **languages**—over Σ.

Thus, this definition is very general: A formal language need not have any form whatsoever! The reader might also find our terminology somewhat unusual in general. A language should consist of sentences rather than of words, as is the case in our terminology. However, as already pointed out above, this is irrelevant and depends merely on the choice of the basic terminology; we have chosen the "neutral" mathematical terminology.

For instance,

$$L_1 = \{\lambda, 0, 010, 1110\} \quad \text{and} \quad L_2 = \{0^p \mid p \text{ prime}\}$$

are languages over the alphabet $\Sigma = \{0, 1\}$, the former being finite and the latter infinite. Here, L_2 is also a language over the alphabet $\Sigma_1 = \{0\}$. In general, if L is a language over the alphabet Σ_1 and Σ is an alphabet containing Σ_1, then L is also a language over Σ. However, when we speak of the *alphabet of a language L*, denoted by ALPH(L), then we mean the smallest alphabet Σ such that L is a language over Σ. Thus, $\text{ALPH}(L_1) = \{0, 1\}$ and $\text{ALPH}(L_2) = \{0\}$. If L consists of a single word, $L = \{w\}$, then we write simply $\text{ALPH}(w)$ instead of $\text{ALPH}(\{w\})$. In general, we identify elements x and singleton sets $\{x\}$ whenever there is no danger of confusion.

Specific families of languages are often conveniently characterized in terms of operations defined for languages: The family consists of all languages obtainable from certain given languages by certain operations. We now define some of the most-common operations. Others will be defined later on.

Regarding languages as sets, we may immediately define the **Boolean operations** of union, intersection, complementation, and difference in the natural fashion. The customary notations $L \cup L'$, $L \cap L'$, $\sim L$ and $L - L'$ are used. In defining the complement of L, $\sim L$, we often consider $\text{ALPH}(L) = \Sigma$: $\sim L$ consists of all words in Σ^* that are not in L. Thus,

$$\sim L = \Sigma^* - L.$$

(This is done in order to avoid any ambiguity in the definition of complementation. When defining the other Boolean operations, the alphabet need not be considered. One should, however, be careful; if complement is defined using ALPH, then some of the customary formulas are not necessarily valid. An example of such a formula is $\sim\sim L = L$.)

The **catenation** (or *product*) **of two languages** L_1 and L_2 is defined by

$$L_1 L_2 = \{w_1 w_2 \mid w_1 \in L_1 \quad \text{and} \quad w_2 \in L_2\}.$$

The notation L^i is extended to apply to the catenation of languages. By definition, $L^0 = (\lambda)$. Observe that this definition guarantees that the customary equations

$$L^i L^j = L^{i+j} \quad \text{and} \quad (L^i)^j = L^{ij}$$

hold for all languages L and nonnegative integers i and j. Observe also that the empty language, $\varnothing$, is not the same as the language $\{\lambda\}$. Indeed, $\varnothing$ and $\{\lambda\}$ can be considered as zero and unit elements with respect to catenation because, for any language L,

$$L\varnothing = \varnothing L = \varnothing, \qquad L\{\lambda\} = \{\lambda\}L = L.$$

The **catenation closure** of a language L, L^*, is defined to be the union of all powers of L:

$$L^* = \bigcup_{i=0}^{\infty} L^i.$$

The **λ-free catenation closure** of L, L^+, is defined to be the union of all positive powers of L:

$$L^+ = \bigcup_{i=1}^{\infty} L^i.$$

Thus, a word is in L^+ iff it is obtained by catenating a finite number of words belonging to L. The empty word, λ, is in L^* for every L (including $L = \varnothing$) because $L^0 = \{\lambda\}$. Observe also that the notations Σ^* and Σ^+ introduced previously are in accordance with the definition of the operations L^* and L^+ if Σ is viewed as the finite language consisting of all single-letter words. For instance,

$$\{a^{2n} \mid n \geq 1\} = \{a^2\}^+ \quad \text{and} \quad \{a^{7n+3} \mid n \geq 0\} = \{a^7\}^*\{a^3\}.$$

An operation of crucial importance in language theory is the operation of morphism. A mapping $h : \Sigma^* \rightarrow \Delta^*$, where Σ and Δ are alphabets, satisfying the condition

$$h(ww') = h(w)h(w'), \qquad \text{for all words } w \text{ and } w' \tag{1}$$

is called a **morphism.** For languages L over Σ, we define

$$h(L) = \{h(w) \mid w \text{ is in } L\}.$$

(Again, algebraically speaking, a morphism of languages is a monoid morphism linearly extended to subsets of monoids.) In view of the condition in (1), to define a morphism h, it suffices to list all the words $h(a)$, where a ranges over all the finitely many letters of Σ. A morphism h is called **nonerasing** (resp. **letter-to-letter**) if $h(a) \neq \lambda$ (resp. $h(a)$ is a letter) for every a in Σ.

We have pointed out that a finite language can be defined, at least in principle, by listing all the words in it, whereas such a definition is not possible for infinite languages. We have already seen how to define infinite languages by specifying a property that must be satisfied by the words in the language. An example is the language $\{0^p \mid p \text{ prime}\}$. The operations introduced above give a way of defining infinite languages because each of the operations $\sim$, $*$, and $^+$ yields an infinite language when applied to a finite language containing at least one nonempty word. For instance, we may consider all languages obtainable from the *atomic* languages $\varnothing$ and $\{a\}$, where a ranges over the letters of some alphabet Σ, by finitely many applications of the operations introduced above. Such languages are called *regular* in Chapter 3, where it will be also seen that we need only a few of the operations introduced above to get all these languages.

We shall introduce a general model for the definition of languages by means of "legal" derivations. The model is referred to as a **rewriting system.** The notion of a (phrase-structure) **grammar** is obtainable from this model by providing it with an input and output format. Before introducing this model, we still want to consider four examples of a somewhat more sophisticated nature than the examples mentioned above. The first three examples deal with operations and are also of general theoretical interest: Example 2.1 in regard to operations in general, Example 2.2 for regular languages, and Example 2.3 for cryptography. The fourth example introduces the notion of a rewriting system.

Example 2.1. Consider the language L over the alphabet $\{a, b, c\}$ consisting of all words of the form

$$c^i w c^j, \quad i \geq 0, j \geq 0,$$

where w is the empty word, the letter a is a prefix of w, or the letter b is a suffix of w. (Thus, for instance, λ, c^3, $cacbac^2$, ca, and bc are all in L, whereas none of the words ba, $c^3 bca^3 c$, $c^2 bc^7 a$ is in L.) Although L misses many words over the alphabet $\{a, b, c\}$, we claim that

$$L^2 = \{a, b, c\}^*. \tag{2}$$

Consequently, since λ is in L, $L^i = \{a, b, c\}^*$ for every $i \geq 2$.

To establish the claim in (2), we prove that an arbitrary given word x over $\{a, b, c\}$ is in L^2. This is obvious if $x = \lambda$. If a is a prefix of x, then x is in L and, hence, also in L^2. (Observe that L^2 contains L.) If b is a prefix of x, we may write x in the form $x = bz$ or $x = bybz$, for some words y and z such that b is not in ALPH(z). Clearly, the words b, byb, and z are in L and, consequently, x is in L^2. Finally, let c be a prefix of x. If b is not in ALPH(x), then clearly x is in L. Otherwise, we may write x in the form

$$x = c^i ybz,$$

for some $i \geq 0$ and words y and z such that b is not in ALPH(z). Again, both $c^i yb$ and z are in L and, consequently, x is in L^2. Since we have exhausted all cases, the claim in (2) follows. The reader might want to prove (2) by considering the cases: a occurs in x and a does not occur in x.

Example 2.2. Define the language L by $L = \{ababa\}^*$. Thus, L consists of the empty word λ and of all words of the form $(ababa)^n$, where $n \geq 1$. We want to show that L can be obtained from the atomic languages $\varnothing$, $\{a\}$, and $\{b\}$ without using the star operation. (The definition above shows how L is obtained from the atomic languages by the operations of star and catenation.) We claim that L can be obtained from the atomic languages by the operations of catenation, union, and complementation.

Let $\Sigma = \{a, b\}$ and observe that $\sim\varnothing = \Sigma^*$. Observe also that intersection can be expressed in terms of union and complementation:

$$L_1 \cap L_2 = \sim(\sim L_1 \cup \sim L_2)$$

for all languages L_1 and L_2. Finally, observe that

$$\{\lambda\} = \sim((\{a\} \cup \{b\})\Sigma^*).$$

Consequently, we may use each of the items Σ^*, $\cap$, and $\{\lambda\}$ without loss of generality in our following considerations.

Since the nonempty words in L are $ababa$, $ababaababa$, $ababaababaababa,\ldots$, we conclude that the words

$$ababa, \qquad babaa, \qquad abaab, \qquad baaba, \qquad aabab \tag{3}$$

are the only words of length 5 appearing as a subword in some word in L. (Indeed, this conclusion can be made by considering only the word *ababaababa*, since any word of length 5 appearing as a subword in some word in L must appear as a subword of *ababaababa*.) There are altogether 32 words of length 5 over the alphabet $\{a, b\}$. Let $w_1, w_2, \ldots, w_{27}$ be all those words of length 5 that are not among the words in (3).

To establish our claim, it suffices to prove that

$$L = \{\lambda\} \cup [(ababa)\Sigma^* \cap \Sigma^*(ababa) \cap \sim (\Sigma^* w_1 \Sigma^*) \cap \cdots \cap \sim (\Sigma^* w_{27} \Sigma^*)] \quad (4)$$

(We have already observed that $\{\lambda\}$, Σ^* and $\cap$ can be expressed in terms of union, complementation, and catenation.) Since every nonempty word in L has the word *ababa* both as a prefix and as a suffix and none of the words $w_1, \ldots, w_{27}$ as a subword, we conclude that the left side of (4) is included in the right side. To prove the reverse inclusion, we assume the contrary and let x be the shortest word that is in the language of the right side of (4) but not in L. The word x must be nonempty because λ is in L. Consequently, x belongs to each of the languages listed within brackets on the right side of (4). Since x belongs to the first of these languages, we may write $x = ababa\, x_1$ for some word x_1. Here x_1 is not in L because otherwise x is in L. This implies that x_1 is also not in the language of the right side of (4) because otherwise we have a contradiction with our choice of x as the *shortest* word in the difference between the right and left sides of (4).

Consequently, x_1 does not belong to all the languages listed within brackets. On the other hand, it must belong to all except $(ababa)\Sigma^*$. For if x_1 has a wrong suffix or a wrong subword, so does x. (The case $|x_1| < 5$ is easily taken care of.) Hence, we conclude that x_1 is not in $(ababa)\Sigma^*$.

Clearly, $x_1 \neq \lambda$ because λ is in L. Hence, x_1 has either a or b as a prefix. If it has b as a prefix, then x has the wrong subword *babab*. Therefore, $x_1 = ax_2$ for some x_2. Moreover, $x_2 \neq \lambda$ because otherwise x has the word *aa* as a suffix.

In the same way we see now that $x_2 = baba\, x_3$ for some x_3 because otherwise x has one of the forbidden words *abaaa*, *baabb*, *aabaa*, or *ababb* (respectively, when proceeding from left to right) as a subword, or else x is not in $\Sigma^*(ababa)$. But this implies that, after all, x_1 is in $(ababa)\Sigma^*$, which is a contradiction.

Somewhat shorter subwords can be used in the above argument. On the other hand, the length 5 is the natural one for consideration concerning our language L. Observe also that $\varnothing$ can be defined by $\varnothing = \{a\} \cap \{b\}$.

The above construction does not work for the slightly modified language $L' = \{abab\}^*$: For L', the star operation is quite essential and can not be avoided, as above. The reader might want to prove this and find reasons for this state of affairs.

Example 2.3.   Consider the English alphabet $\Sigma = \{A, B, \ldots, Z\}$. Each of the letter-to-letter morphisms

TABLE 2.1

	A	B	C	D	E	F	G	H	I	J	K	L	M	N	O	P	Q	R	S	T	U	V	W	X	Y	Z
0	A	B	C	D	E	F	G	H	I	J	K	L	M	N	O	P	Q	R	S	T	U	V	W	X	Y	Z
1	B	C	D	E	F	G	H	I	J	K	L	M	N	O	P	Q	R	S	T	U	V	W	X	Y	Z	A
2	C	D	E	F	G	H	I	J	K	L	M	N	O	P	Q	R	S	T	U	V	W	X	Y	Z	A	B
3	D	E	F	G	H	I	J	K	L	M	N	O	P	Q	R	S	T	U	V	W	X	Y	Z	A	B	C
4	E	F	G	H	I	J	K	L	M	N	O	P	Q	R	S	T	U	V	W	X	Y	Z	A	B	C	D
5	F	G	H	I	J	K	L	M	N	O	P	Q	R	S	T	U	V	W	X	Y	Z	A	B	C	D	E
6	G	H	I	J	K	L	M	N	O	P	Q	R	S	T	U	V	W	X	Y	Z	A	B	C	D	E	F
7	H	I	J	K	L	M	N	O	P	Q	R	S	T	U	V	W	X	Y	Z	A	B	C	D	E	F	G
8	I	J	K	L	M	N	O	P	Q	R	S	T	U	V	W	X	Y	Z	A	B	C	D	E	F	G	H
9	J	K	L	M	N	O	P	Q	R	S	T	U	V	W	X	Y	Z	A	B	C	D	E	F	G	H	I
10	K	L	M	N	O	P	Q	R	S	T	U	V	W	X	Y	Z	A	B	C	D	E	F	G	H	I	J
11	L	M	N	O	P	Q	R	S	T	U	V	W	X	Y	Z	A	B	C	D	E	F	G	H	I	J	K
12	M	N	O	P	Q	R	S	T	U	V	W	X	Y	Z	A	B	C	D	E	F	G	H	I	J	K	L
13	N	O	P	Q	R	S	T	U	V	W	X	Y	Z	A	B	C	D	E	F	G	H	I	J	K	L	M
14	O	P	Q	R	S	T	U	V	W	X	Y	Z	A	B	C	D	E	F	G	H	I	J	K	L	M	N
15	P	Q	R	S	T	U	V	W	X	Y	Z	A	B	C	D	E	F	G	H	I	J	K	L	M	N	O
16	Q	R	S	T	U	V	W	X	Y	Z	A	B	C	D	E	F	G	H	I	J	K	L	M	N	O	P
17	R	S	T	U	V	W	X	Y	Z	A	B	C	D	E	F	G	H	I	J	K	L	M	N	O	P	Q
18	S	T	U	V	W	X	Y	Z	A	B	C	D	E	F	G	H	I	J	K	L	M	N	O	P	Q	R
19	T	U	V	W	X	Y	Z	A	B	C	D	E	F	G	H	I	J	K	L	M	N	O	P	Q	R	S
20	U	V	W	X	Y	Z	A	B	C	D	E	F	G	H	I	J	K	L	M	N	O	P	Q	R	S	T
21	V	W	X	Y	Z	A	B	C	D	E	F	G	H	I	J	K	L	M	N	O	P	Q	R	S	T	U
22	W	X	Y	Z	A	B	C	D	E	F	G	H	I	J	K	L	M	N	O	P	Q	R	S	T	U	V
23	X	Y	Z	A	B	C	D	E	F	G	H	I	J	K	L	M	N	O	P	Q	R	S	T	U	V	W
24	Y	Z	A	B	C	D	E	F	G	H	I	J	K	L	M	N	O	P	Q	R	S	T	U	V	W	X
25	Z	A	B	C	D	E	F	G	H	I	J	K	L	M	N	O	P	Q	R	S	T	U	V	W	X	Y

$$h_i : \Sigma^* \to \Sigma^*, \qquad 0 \leq i \leq 25,$$

maps each letter to the letter lying i positions further in the alphabet, whereby the end of the alphabet is continued cyclically to the beginning. Thus,

$$h_2(A) = C, \qquad h_7(Y) = F, \qquad h_{25}(Z) = Y.$$

In cryptography, each of the morphisms h_i is customarily referred to as a **Caesar cipher.**

Table 2.1, usually referred to as the **Vigenère table,** gives the definition of h_i in its ith row, for each $i = 0, 1, \ldots, 25$. The table is used for encryptions based on the Caesar ciphers h_i.

Here are some further illustrations:

$$h_{25}(IBM) = HAL, \qquad h_3(HELP) = KHOS.$$

Observe that the morphisms h commute:

$$h_i h_j = h_j h_i \qquad \text{for every } i \text{ and } j.$$

(Here the juxtaposition stands for the *composition* of morphisms. In general, if g and h are morphisms, then by gh we mean the morphism obtained by applying first h and then g to the result. Of course, in order that the composition be defined, the range and domain alphabets must match.) Moreover,

$$h_{26-i} h_i = h_0 \qquad \text{for every } i \geq 1.$$

Thus, if some plaintext is encrypted using h_i, the same plaintext can be recovered by decrypting the ciphertext using h_{26-i}.

Example 2.4. The language L over the alphabet $\{a, b\}$ consists of all words that can be obtained from the empty word λ by finitely many applications of the following three rules.

> **Rule 1.** If w is in L, then so is awb.
> **Rule 2.** If w is in L, then so is bwa.
> **Rule 3.** If w_1 and w_2 are in L, then $w_1 w_2$ is also in L.

For instance, Rule 1 shows that ab is in L. Hence, by Rule 2, $baba$ is in L and, by Rule 3, $babaab$ is also in L. This same word can also be derived by first applying Rules 2 and 1 to get ba and ab and then applying Rule 3 twice. Hence, the derivation of a word in L is by no means unique.

We claim that L consists of all words with an equal number of a's and b's. Indeed, since λ has this property and Rules 1–3 preserve it, we conclude that L contains only words with this property. We still have to show that, conversely, every word with this property is in L.

Clearly, L contains all words of length 0 with equally many a's and b's. Proceeding inductively, we assume that L contains all words of length no greater than $2i$ with this property. Let w be an arbitrary word of length $2(i+1)$ with this property. Clearly $w \neq \lambda$. We assume without loss of generality that the first letter of w is a. (If this is not the case originally, we may interchange the roles of a and b because Rules 1–3 are symmetric with respect to a and b.) If the last letter of w is b, we obtain $w = aw_1b$, where w_1 is a word of length $2i$ with equally many a's and b's. By our inductive hypothesis, w_1 is in L. Hence, by Rule 1, w is in L. If the last letter of w is a, w must have a nonempty prefix $w_1 \neq w$ containing equally many a's and b's. Consequently, $w = w_1w_2$, where both w_1 and w_2 are words of length no greater than $2i$ with equally many a's and b's. By our inductive hypothesis, both w_1 and w_2 are in L. Hence, by Rule 3, w is in L also in this case. $\square$

The rules in the above example permit certain rewriting to take place. For instance, w can be rewritten as awb. The following definition captures the essence of this rewriting procedure. A rewriting rule can also be viewed as an abstraction of a rule of inference in the sense customary in logic.

Definition 2.1. *A rewriting system is a pair $RW = (\Sigma, P)$, where Σ is an alphabet and P is a finite set of ordered pairs of words over Σ. The elements (u, w) of P are referred to as* **rewriting rules,** *or* **productions,** *and are denoted by $u \rightarrow w$.*

Given a rewriting system RW, the **yield relation** *$\Rightarrow_{RW}$, or $\Rightarrow$ if there is no danger of confusion, is defined as follows. For any x and y in Σ^*, $x \Rightarrow y$ holds if there are words x_1 and x_2 such that*

$$x = x_1ux_2, \qquad y = x_1wx_2,$$

for some production $u \rightarrow w$ in the system. We say that x **yields directly** *y according to RW.*

The word x **yields** *the word y, denoted by $x \Rightarrow *_{RW} y$, or $x \Rightarrow *y$, iff there is a finite sequence of words over Σ*

$$w_0, w_1, \ldots, w_k, \qquad k \geq 0,$$

where $w_0 = x$, $w_k = y$, and $w_i \Rightarrow w_{i+1}$, for $0 \leq i \leq k - 1$. This sequence is referred to as a **derivation** *of y from x according to RW, k being the* **length** *of the derivation. Derivations are also written*

$$w_0 \Rightarrow w_1 \Rightarrow \cdots \Rightarrow w_k.$$ $\square$

The binary relation $\Rightarrow *$ can also be viewed as the **reflexive transitive closure** of the binary relation $\Rightarrow$. In general, if ρ is a binary relation on a set S, then the reflexive transitive closure $\rho *$ of ρ is defined as follows:

(i) $s\rho *s$ holds for every s in S.
(ii) If $s_1\rho *s_2$ and $s_2\rho s_3$, then $s_1\rho *s_3$.
(iii) $s\rho *s'$ only if it can be established by (i) and (ii).

A rewriting system can be converted into a language-defining device in various ways. One may choose a finite set of words ("axioms") and consider the language consisting of all words derivable from some of the axioms according to the system. Or, one may consider the language consisting of all words w such that one of the axioms is derivable from w. The alphabet Σ may also contain certain auxiliary (also referred to as nonterminal) letters; words containing auxiliary letters are not included in the language considered.

We do not want to formalize these ideas for general rewriting systems. However, the next section contains a specific formalization. To illustrate the ideas, we return to Example 2.4.

Consider the rewriting system

$$RW = (\{a, b, S\}, \{S \to \lambda, S \to aSb, S \to bSa, S \to SS\}).$$

Associate with RW the language

$$L_{RW} = \{w \in \{a, b\}^* \mid S \Rightarrow {}^*_{RW} w\}.$$

Thus, RW is augmented by choosing S to be the (only) axiom and by also considering S as a nonterminal letter; only words over the alphabet $\{a, b\}$ are included in the language L_{RW}. Some production must still be applied to words containing S in order to get a word in L_{RW}.

It is now easy to verify that $L_{RW} = L$, where L is the language considered in Example 2.4. Observe, in particular, the role of the productions: The first production corresponds to the starting point λ, whereas the purpose of the other three productions is identical to that of Rules 1–3.

A rewriting system is the basic model for *partial* rewriting; at each step of the rewriting process only some part of the word is rewritten, whereas the other parts remain unchanged. This is to be contrasted with *parallel* rewriting, where at each step of the rewriting process all letters of the word considered have to be rewritten. A brief discussion of parallel rewriting is given in Section 2.5. One may also introduce a more-general model, containing the rewriting systems introduced above, as well as parallel rewriting systems as special cases. Such a model has an additional parameter specifying the mode of rewriting.

2.2. GRAMMARS

The notion of a phrase structure grammar was originally introduced for linguistic applications. The nonterminals represent various syntactic classes of a natural language. Terminals represent words, and the productions capture the essence of the parsing process. However, the same notion can be applied for programming languages as well, and also for various considerations dealing with the theory of computing in general.

Definition 2.2. *A **phrase-structure grammar**, or, briefly, a **grammar**, is a quadruple*

$$G = (RW, \Sigma_N, \Sigma_T, S),$$

*where $RW = (\Sigma, P)$ is a rewriting system, Σ_N and Σ_T are disjoint alphabets such that $\Sigma = \Sigma_N \cup \Sigma_T$, and S is an element of Σ_N. The elements of Σ_N and Σ_T are referred to as **nonterminal** and **terminal** letters, respectively. S is called the **start letter**. The language **generated** by the grammar G is defined by*

$$L(G) = \{w \in \Sigma_T^* \mid S \underset{RW}{\overset{*}{\Rightarrow}} w\}.$$

*Two grammars are **equivalent** iff they generate the same language.*   □

Thus, a grammar can be defined by specifying the two alphabets Σ_N and Σ_T and a special start letter from the former alphabet, as well as a finite set of productions—that is, ordered pairs of words over the union of the two alphabets. We usually define a grammar simply by listing the productions and applying the *convention* that capital letters are nonterminals and lower-case letters are terminals. Unless stated otherwise, S (possibly provided with some indices) will be the start letter.

For instance, consider the grammar G_1 defined by the productions

$$S \to aSb, \qquad S \to ab,$$

the grammar G_2 defined by the productions

$$S \to aSb, \qquad S \to SS, \qquad S \to \lambda,$$

and the grammar G_3 having the productions

$$S \to aSb, \qquad S \to bSa, \qquad S \to SS, \qquad S \to \lambda.$$

A typical derivation according to G_2 is the following:

$$S \Rightarrow SS \Rightarrow aSbS \Rightarrow aaSbbS \Rightarrow aabbS \Rightarrow aabbaSb \Rightarrow aabbab.$$

Thus, the word *aabbab* is in $L(G_2)$.

It is easy to verify that

$$L(G_1) = \{a^n b^n \mid n \geq 1\}.$$

The language $L(G_3)$ equals the language L considered in Example 2.4 above. (Observe again how the productions correspond to Rules 1–3.) Finally, the language $L(G_2)$ is a subset of $L(G_3)$, consisting of all *properly nested* parentheses when a is viewed as the left parenthesis and b as the right parenthesis. In this notation, the word derived above becomes (( ))( ), indeed a properly nested sequence of parentheses. The word *abba* of $L(G_3)$ corresponds to the improper nesting ( ))(and, hence, cannot be derived according to G_2.

As already pointed out, nonterminals were originally intended to represent syntactic classes of a natural language. Consider, for instance, a grammar with the nonterminals S, Np, N, Vp, V, A (corresponding to the syntac-

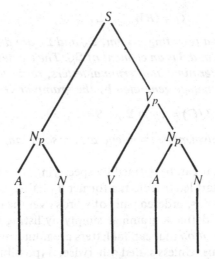

the computer swallowed the information

FIGURE 2.1.

tic classes *sentence, noun phrase, noun, verb phrase, verb,* and *article,*
respectively). Natural productions would then be:

$$S \rightarrow Np\,Vp, \qquad Np \rightarrow AN, \qquad Vp \rightarrow VNp.$$

(A sentence consists of a noun phrase followed by a verb phrase, and so on.)
If we still add these productions involving terminals

$$N \rightarrow \text{computer}, \qquad N \rightarrow \text{information}, \qquad A \rightarrow \text{the}, \qquad V \rightarrow \text{swallowed},$$

we get, for instance, the derivation

$$S \Rightarrow Np\,Vp \Rightarrow ANVp \Rightarrow ANVNp \Rightarrow ANVAN \Rightarrow *$$

The computer swallowed the information.

The derivation can be illustrated by the *derivation tree,* or *parse tree* in Figure 2.1. Of course, from the point of view of actual parsing, it is more natural to view the productions in the reverse way: from right to left. This means
that the above tree is processed in a bottom-up rather than top-down fashion.

Returning to the general notion of a grammar, we next give two somewhat more sophisticated examples.

Example 2.5.   Consider the grammar G determined by the productions

$$S \rightarrow abc, \qquad S \rightarrow aAbc, \qquad Ab \rightarrow bA, \qquad Ac \rightarrow Bbcc, \qquad bB \rightarrow Bb,$$
$$aB \rightarrow aaA, \qquad aB \rightarrow aa.$$

We claim that

$$L(G) = \{a^n b^n c^n \mid n \geq 1\}. \tag{5}$$

In fact, any derivation according to G begins with an application of the first or second production, the first production directly yielding the terminal word abc. Consider any derivation D from a word $a^i A b^i c^i$, where $i \geq 1$, leading to a word over the terminal alphabet. D must begin with i applications of the third production (A travels to the right) and then continue with an application of the fourth production (one further occurrence of b and c is deposited). Now we have derived the word $a^i b^i B b c^{i+1}$. For this the only possibility is to apply the fifth production i times (B travels to the left), after which we have the word $a^i B b^{i+1} c^{i+1}$. This word directly yields one of the two words

$$a^{i+1} A b^{i+1} c^{i+1} \quad \text{or} \quad a^{i+1} b^{i+1} c^{i+1}$$

by the last two productions (one further a is deposited, and either a new cycle is entered or the derivation is terminated). This argument shows that G generates all words belonging to the right side of (5) and nothing else; every step in a derivation is uniquely determined, the only exception being that there is a choice between termination and entrance into a new cycle.

The reader may verify that the grammar

$$G_1: \quad S \to aSAB, \quad S \to abB, \quad BA \to AB, \quad bA \to bb, \quad bB \to bc,$$
$$cB \to cc,$$

and the grammar

$$G_2: \quad S \to aSA, \quad S \to aB, \quad BA \to bBc, \quad cA \to Ac, \quad B \to bc$$

are both equivalent to G.

Example 2.6. Define the **mirror image of a word** w, denoted by $\mathrm{mi}(w)$, to be the word obtained by writing w backwards. For instance,

$$\mathrm{mi}(ababb) = bbaba.$$

By definition, $\mathrm{mi}(a) = a$ for a letter a and $\mathrm{mi}(\lambda) = \lambda$. The **mirror image of a language** L is defined by

$$\mathrm{mi}(L) = \{\mathrm{mi}(w) \mid w \text{ in } L\}.$$

Consider now the following two languages:

$$L_1 = \{w\mathrm{mi}(w) \mid w \in \{a, b\}^*\},$$
$$L_2 = \{ww \mid w \in \{a, b\}^*\}.$$

Although L_1 and L_2 look equally complex, it is much simpler to define the language L_1 by a grammar. In fact, L_1 is generated by the following grammar, G_1:

$$S \to \lambda, \qquad S \to aSa, \qquad S \to bSb.$$

To generate L_2, we consider the following grammar, G_2:

$$S \to ABC, \tag{6}$$

$$AB \to iAD_i, \qquad D_i j \to jD_i, \qquad D_i C \to BiC, \qquad iB \to Bi, \tag{7}$$

$$AB \to \lambda, \qquad C \to \lambda, \tag{8}$$

where i and j independently assume the values a and b.

Consider derivations from $wABwC$ leading to a terminal word. (After an application of the initial production in (6), we have $w = \lambda$.) If one of the productions of (8) is used, then the other must also be used, and the word ww results. If our derivation begins with an application of the first production in (7), then the only possibility is to continue the derivation to the words $wiAwD_iC$ and $wiAwBiC$. From the latter, the only possibility is to generate one of the words $wiABwiC$ or $wiABwi$. The former is of our original form, and the latter necessarily leads to the word $wiwi$. Inductively, we conclude that $L(G_2) = L_2$. □

The terminology we now introduce will be very important throughout all our following considerations dealing with languages.

We say that a language is **recursively enumerable** iff it is generated by a grammar. The motivation behind this term will become clearer in Chapter 4. However, it is obvious by the very definition of a grammar and its generated language that there is an effective procedure of enumerating (that is, listing) all words in the language generated by a grammar. To accomplish this, one simply lists all derivations of length 1, and after that all derivations of length 2, and so on. Whenever a word w over the terminal alphabet appears as the last word of a derivation, then w is added to our listing of the language. Because for any k, there clearly are only finitely many derivations of length k, the procedure described provides an effective listing (with eventual repetitions) of the language generated by our grammar.

A grammar is termed **context-free** iff the left side of every production consists of a single nonterminal letter. Similarly, a language is termed context-free iff it is generated by a context-free grammar.

Intuitively, the term *context-free* means that rewriting is always independent of the context. If $A \to \alpha$ is one of the productions, then the nonterminal A can be rewritten as α, no matter what the letters adjacent to A are. This is to be contrasted with **context-sensitive** rewriting. For instance, if we have the two productions $BAC \to B\alpha C$ and $DAE \to D\beta E$, we may rewrite an occurrence of A between B and C as α and an occurrence of A between D and E as β—but not vice versa!

For instance, the grammar G_1 considered in Example 2.6 is context-free, and so are all the grammars considered in this section before Example

2.5. On the other hand, none of the grammars considered in Example 2.5 is context-free. In fact, it can be shown that the language in (5) is not generated by any context-free grammar. A useful tool for this argument will be provided in the next chapter.

A context-free grammar is called **regular** iff every production is of the form $A \to wB$ or of the form $A \to w$, where A and B are nonterminal letters and w is a word (including the empty word) over the terminal alphabet. Similarly, a language is regular iff it is generated by a regular grammar.

Any detailed theory of context-free and regular languages is beyond the scope of this book. The reader is referred to [Sa3]. However, some facts will be needed in the forthcoming chapters. This section is concluded with three useful "normal form" results. The results show that every language of a certain type (recursively enumerable, context-free, regular) can be generated by a grammar that is simpler than the general grammar of this type given by the definition. We consider first regular, then context-free, and, finally, recursively enumerable languages.

Theorem 2.1. *Every regular language is generated by a grammar such that every production is of the form $A \to aB$ or of the form $A \to \lambda$, where A and B are nonterminal letters and a is a terminal letter. Moreover, for each pair (A, a), there is exactly one production $A \to aB$ in the grammar.*

Proof. Assume that $L = L(G)$, where G is a regular grammar. We will construct a grammar G_1 equivalent to G such that all productions of G_1 are of the two legal forms mentioned in the statement of Theorem 2.1. In addition, G may originally contain productions of the following forms:

$$A \to a_1 \cdots a_k B, \qquad k \geq 2, a_j \in \Sigma_T; A, B \in \Sigma_N, \tag{9}$$

$$A \to a_1 \cdots a_k, \qquad k \geq 1, a_j \in \Sigma_T, A \in \Sigma_N, \tag{10}$$

$$A \to B, \qquad A, B \in \Sigma_N. \tag{11}$$

Thus, we have to eliminate all productions in (9)–(11) from G.

For each production in (9), we introduce new nonterminals B_1, $\ldots$, B_{k-1} and replace (9) by the productions

$$A \to a_1 B_1, B_1 \to a_2 B_2, \ldots, B_{k-1} \to a_k B.$$

For each production in (10), we introduce new nonterminals $A_1, \ldots, A_k$ and replace (10) by the productions

$$A \to a_1 A_1, A_1 \to a_2 A_2, \ldots, A_{k-1} \to a_k A_k, A_k \to \lambda.$$

This procedure is repeated for each production in (9) and (10). (It is important that the additional nonterminals are always new ones: No additional nonterminal is used in connection with two different productions.) It is immediate that the language $L(G)$ is not affected by these replacements. At the same time, the "wrong" forms of (9) and (10) are eliminated. We still have to eliminate productions (11)—without, of course, creating new productions of a wrong form!

We proceed as follows. Let A be an arbitrary nonterminal. (We are now considering the collection of all nonterminals in our grammar, after the elimination of productions (9) and (10).) Define the following sets of nonterminals:

$$N_1(A) = \{A\},$$
$$N_{i+1}(A) = N_i(A) \cup \{B \mid \text{for some } C \in N_i(A), C \to B \text{ is a production}\}.$$

Since there are altogether only finitely many nonterminals, there is a k with the property $N_k(A) = N_{k+1}(A)$. It is easy to verify that, for this k, we have $N_k(A) = N_{k+i}(A)$, $i = 1,2,\ldots$, and furthermore, for an arbitrary nonterminal B,

$$A \Rightarrow^* B \quad \text{iff} \quad B \in N_k(A).$$

We have, thus, given a method of determining, for an arbitrary nonterminal A, the set $N(A)$ consisting of all nonterminals B with the property $A \Rightarrow^* B$.

We now remove all productions (11) from our grammar. For each production $A \to aB$, we add to the grammar every production $A \to aC$ such that C is in $N(B)$. Moreover, we add to the grammar a new start letter $\overline{S}$ and all productions $\overline{S} \to aC$ such that, before removing the productions (11), our grammar had the production $S' \to aB$, where $C \in N(B)$ and $S' \in N(S)$ with S being the previous start letter. Also, the production $\overline{S} \to \lambda$ is added if necessary—that is, if λ is in the language $L(G)$. (Clearly, whether or not λ is in $L(G)$ can be immediately tested.)

It is easy to verify that $L(G)$ is not affected by these changes. Indeed, the applications of "chain" productions $A \to B$ are taken care of by our new productions: If we originally first applied the production $A \to aB$ and then a sequence of chain productions, we can get the same effect in the new grammar by just one single application of the production $A \to aC$, where C is in $N(B)$. Moreover, the initial applications of chain productions (for instance, $S \Rightarrow S_1 \Rightarrow S_2 \Rightarrow S_3$) have been implemented in the productions for the new start symbol $\overline{S}$.

We have, thus, eliminated all productions in (9)–(11). To complete the proof, we still have to modify our grammar to satisfy the uniqueness condition.

The terminal alphabet remains unchanged, but the new nonterminals will be subsets of the set of our previous nonterminals, including the empty subset. Thus, if our nonterminals before were $\overline{S}$, A, and B, they are now

$$[\varnothing], [\overline{S}], [A], [B], [\overline{S}, A], [\overline{S}, B], [A, B], [\overline{S}, A, B].$$

Brackets are used to indicate that the subset is viewed as one letter. In general, if there were n nonterminals in the previous grammar, there are 2^n of them in the new grammar.

The new start letter is the subset $[\bar{S}]$ consisting of the old start letter alone. The productions are defined as follows:

(i) For a subset X of nonterminals, the production $[X] \to \lambda$ is in the new grammar iff there is a nonterminal A in X such that the production $A \to \lambda$ is in the previous grammar. (Thus, the production $[\varnothing] \to \lambda$ is never in the new grammar.)

(ii) For a subset X of nonterminals and a terminal letter a, the production $[X] \to a[Y]$ is in the new grammar, where Y is the unique subset of nonterminals defined by

$Y = \{B \mid$ for some A in X, the production $A \to aB$ is in the previous grammar$\}$.

(Thus, for all a, $[\varnothing] \to a[\varnothing]$ is a production in the new grammar.)

This completes the definition of the new grammar G_1. It is immediate that G_1 satisfies the uniqueness condition mentioned in the second sentence of Theorem 2.1. It is also easy to verify that G_1 generates the same language as the previous grammar; our subset construction is capable of simulating exactly the derivations of the previous grammar. $\square$

Theorem 2.2.   Every λ-free (that is, not containing the empty word) context-free language is generated by a grammar such that every production is of the form $A \to BC$ or of the form $A \to a$, where A, B, and C are nonterminal letters and a is a terminal letter.

Proof.   The proof is similar to the previous proof: we eliminate step by step the wrong productions, all the time keeping the generated language invariant. For this purpose, the removal of wrong productions always has to be compensated by adding productions of the correct form.

We begin by eliminating productions $A \to \lambda$. (Observe that our original grammar generating the language considered may have such productions, although λ is not in the language.) We first determine the set M consisting of all nonterminals B with the property $B \Rightarrow^* \lambda$. (This is done in the same way that the sets $N(A)$ were determined in the previous proof: We first take all the nonterminals directly yielding the empty word and, later on, at each step augment the set by adding every nonterminal directly yielding a word consisting of letters already present.) The new set of productions consists now of all productions $A \to \beta$, $\beta \neq \lambda$, such that there is a production $A \to \alpha$ in the old set of productions and β is obtained from α by deleting 0 or more occurrences of elements of M. Thus, all original productions are preserved with the exception of productions $A \to \lambda$. The effect of the latter is recovered by using the deleted versions of the original productions.

The next step is to get rid of the chain productions $A \to B$. For each letter β, we determine the set $N'(\beta)$ consisting of all terminals and nonterminals α such that $\beta \Rightarrow^* \alpha$. The sets $N'(\beta)$ are determined in almost the same way as the sets $N(A)$ in the preceding proof. The only exception is that we now also have to consider productions whose right side is a single terminal. Recall that the productions $A \to \lambda$ have already been eliminated. Hence, a derivation $A \Rightarrow^* \alpha$ uses length-preserving productions only.

The new set of productions consists now of, first, all productions $S \to a$, where S is the start letter of the grammar and a is a terminal letter in the set $N'(S)$ and, second, all productions $B \to \beta_1 \cdots \beta_n$, where $n \geq 2$ and each β is a letter such that in the old set there is a production $A \to \alpha_1 \cdots \alpha_n$ with the properties

$$A \in N'(B), \qquad \beta_i \in N'(\alpha_i), \qquad i = 1, \ldots, n.$$

This transformation removes the chain productions $A \to B$. It is easy to see that the transformation does not affect the language of our grammar: the chain productions are "hidden" in the new productions involving B's and β's. Observe, in particular, that β itself is always in $N'(\beta)$. So far our transformations have not affected the alphabets of the grammar at all. Thus, there is no danger of confusion in the expression "the start letter of the grammar."

Two more transformations remain. The following step is to make sure that terminal letters appear in productions of the form $A \to a$ only. Consider all productions $A \to \alpha_1 \cdots \alpha_n$, where $n \geq 2$ and the α's are letters. (In addition to such productions, our grammar by now has only productions $S \to a$, which are of the correct form already.) In all such productions we now replace every terminal letter a by a new nonterminal A_a and add the productions $A_a \to a$ to the production set. It is obvious that this transformation has the desired effect.

Finally, for each production

$$A \to B_1 \cdots B_n, \qquad n \geq 3, \tag{12}$$

where the B's are nonterminal letters, we introduce new nonterminals C_1, $\ldots, C_{n-2}$ and replace (12) by the productions

$$A \to B_1 C_1, \, C_1 \to B_2 C_2, \ldots, \, C_{n-3} \to B_{n-2} C_{n-2}, \, C_{n-2} \to B_{n-1} B_n. \qquad \square$$

The normal form given in Theorem 2.2 is called the **Chomsky normal form** for context-free grammars. Of course, a slight modification of the normal form gives also languages containing the empty word.

Theorem 2.3. *Every recursively enumerable language is generated by a grammar such that every production is of one of the forms $A \rightarrow BC$, $AB \rightarrow CD$, $A \rightarrow \lambda$, or $A \rightarrow a$, where A, B, C, and D are nonterminal letters and a is a terminal letter.*

Proof outline. The setup is exactly the same as in the two preceding proofs, and so the reader should be able to fill in the missing details.

One first eliminates terminal letters from productions other than $A \rightarrow a$ by replacing each occurrence of a terminal a with a new nonterminal A_a and adding the productions $A_a \rightarrow a$. The productions $\lambda \rightarrow x$ are replaced by the productions $A \rightarrow Ax$ and $A \rightarrow xA$, where A ranges over all nonterminals. The length-decreasing productions

$$A_1 \cdots A_m \rightarrow B_1 \cdots B_n, \qquad m > n,$$

where the A's and B's are nonterminals, are made length-increasing by adding occurrences of a new nonterminal B to the right side and adding the production $B \rightarrow \lambda$. The productions $A_1 \rightarrow B_1$ and $A_1 \cdots A_m \rightarrow \lambda$ are handled in the same way.

After these transformations, all the remaining wrong productions are of the form

$$A_1 \cdots A_m \rightarrow B_1 \cdots B_n, \qquad m \leq n, n \geq 3, \qquad (13)$$

where A's and B's are nonterminals. If $m = 1$, we proceed exactly as with productions (12) in the preceding proof. If $m \geq 2$, we introduce a new nonterminal C_2 and first replace (13) by the productions

$$A_1 A_2 \rightarrow B_1 C_2, \qquad C_2 A_3 \cdots A_m \rightarrow B_2 \ldots B_n. \qquad (14)$$

(If $m = 2$, then $A_3 \cdots A_m$ reads λ.) The first of the productions in (14) is of the correct form, the second having one symbol less on both sides than (13). If $n > 3$, the same procedure is repeated for the second production (14) until only productions of the correct forms result. $\qquad \square$

The four forms $A \rightarrow BC$, $AB \rightarrow CD$, $A \rightarrow \lambda$, $A \rightarrow a$ in the statement of Theorem 2.3 are independent in the sense that if any of them is omitted, then the resulting statement is not valid any more. Without the first form, we get only languages with words of length less than or equal to 1. Without the second, we get only context-free languages. (In fact, by Theorem 2.2, we get all context-free languages in this fashion.) Without the third form, only context-sensitive languages result (see Exercise 2.6). The fourth form is, of course, necessary for depositing terminals.

In fact, to generate all recursively enumerable languages, you need expansive productions (such as $A \rightarrow BC$), context-sensitive productions (such as

$AB\rightarrow CD$), erasing productions (exemplified by $A\rightarrow\lambda$), and terminating productions (exemplified by $A\rightarrow a$). However, the four requirements mentioned can also be satisfied differently. This gives various modifications of Theorem 2.3. For instance (see Exercise 2.5), productions of the three forms $A\rightarrow BC$, $AB\rightarrow\lambda$, and $A\rightarrow a$ suffice to generate every recursively enumerable language.

2.3. POST SYSTEMS

We now introduce a class of generating devices somewhat different from grammars but still equivalent to grammars in the sense that the family of languages generated by our new devices equals the family of recursively enumerable languages. The new devices—called *Post canonical systems,* or *Post systems,* after Emil Post—very closely resemble formal systems in logic. The productions have the shape of inference rules: Given some premises, a conclusion can be made.

Definition 2.3.   A Post system is a quadruple

$$PS = (\Sigma_V, \Sigma_C, P, S),$$

*where Σ_V and Σ_C are disjoint alphabets, referred to as the **alphabet of variables** and the **alphabet of constants**, respectively; S is a finite set of words over Σ_C, the words being referred to as **axioms**, or **primitive assertions**; and P is a finite set of **productions** where a production is a $(k + 1)$-tuple, $k \geq 1$,*

$$(\alpha_1, \ldots, \alpha_k, \alpha) \tag{15}$$

of words over the alphabet $\Sigma = \Sigma_V \cup \Sigma_C$. For different productions, the number k may be different. $\qquad\square$

Production (15) is usually denoted by $(\alpha_1, \ldots, \alpha_k)\rightarrow\alpha$ (for $k = 1$ we omit the parentheses) and read "the conclusion α can be made from the premises $\alpha_1, \ldots, \alpha_k$." More specifically, the conclusion is made in the following fashion. For all variables appearing in the words $\alpha_1, \ldots, \alpha_k, \alpha$, we uniformly substitute some words over the constant alphabet Σ_C. Here *uniform substitution* means that all occurrences of the same variable A have to be replaced by the same word over the constant alphabet Σ_C. The constant word (meaning: word over Σ_C) $\bar{\alpha}$ obtained in this fashion can be inferred from the corresponding constant words $\bar{\alpha}_1, \ldots, \bar{\alpha}_k$. Initially, we have the constant words in S, and we can use them as premises. Whenever we have inferred some new constant word, we can use it as a premise in later inferences. The language $L(PS, \Sigma_T)$ *generated* by the Post system, PS, with respect to the *terminal* alphabet $\Sigma_T \subseteq \Sigma_C$ consists of all words over Σ_T obtained in this fashion. Thus, we may decompose the alphabet Σ_C in two parts, $\Sigma_C = \Sigma_N \cup \Sigma_T$, such that an inferred word has to be over the alphabet Σ_T in order to be included in the language generated by the Post system. Here Σ_N may be empty.

These ideas are formalized in the following definition.

Definition 2.4. Let $(\alpha_1, \ldots, \alpha_k) \rightarrow \alpha$ *be a production in a Post system, PS. Any production* $(\bar{\alpha}_1, \ldots, \bar{\alpha}_k) \rightarrow \bar{\alpha}$ *is a* **constant version** *of the former production if the latter results from the former by uniformly substituting words over* Σ_C *for letters of* Σ_V. *A word w over* Σ_C *is a* **theorem** *of PS iff either w is an axiom or else there is a constant version* $(\bar{\alpha}_1, \ldots, \bar{\alpha}_k) \rightarrow \bar{\alpha}$ *of some production in PS such that* $\bar{\alpha}_1, \ldots, \bar{\alpha}_k$ *are theorems of PS and* $\bar{\alpha} = w$. *For* $\Sigma_T \subseteq \Sigma_C$, *we define*

$$L(PS, \Sigma_T) = \{w \in \Sigma_T^* \mid w \text{ is a theorem of PS}\}. \qquad \square$$

For instance, the Post system PS with λ as the only axiom, $\Sigma_V = \{A\}$, $\Sigma_C = \{a, b\}$, and with $A \rightarrow aAb$ as the only production (thus, for this production $k = 1$) satisfies

$$L(PS, \Sigma_C) = \{a^n b^n \mid n \geq 0\}. \qquad (16)$$

Thus, in (16) the terminal alphabet chosen equals the whole alphabet, Σ_C. In such a case we say that the language $\{a^n b^n \mid n \geq 0\}$ is generated by a **pure** Post system.

Similarly, the Post system with λ as the only axiom, $\Sigma_V = \{A\}$, $\Sigma_C = \{a, b\}$, and with the productions

$$A \rightarrow aAb, \qquad A \rightarrow bAa, \qquad (A_1, A_2) \rightarrow A_1 A_2$$

generates, viewed as a pure system, the language considered in Example 2.4 above.

The original definition of Post [Po1] had some restrictions for canonical systems. For instance, it was required that every variable appearing in the conclusion α must also appear in one of the premises $\alpha_1, \ldots, \alpha_k$. While such restrictions are natural if the systems are intended to model formal systems of logic, they do not affect the overall generative capacity of the systems. This is basically the reason for not including them in the definition above.

The proofs of the next two theorems are omitted. The first theorem tells that the generative capacity of Post systems is exactly the same as that of grammars. The second theorem is a normal form result.

Theorem 2.4. *A language is recursively enumerable iff it is of the form* $L(PS, \Sigma_T)$, *for some Post system PS and alphabet* Σ_T.

Theorem 2.5.   *For every language $L(PS, \Sigma_T)$, there is a Post system, PS', having productions only of the form $uA \to Aw$, where A is a variable and u and w are words over the alphabet of constants, such that*

$$L(PS, \Sigma_T) = L(PS', \Sigma_T).$$

Consequently, the family of languages generated by Post systems, PS', of this special type equals the family of recursively enumerable languages.

In the remainder of this section we study Post systems of a special type and prove that the languages generated by them are exactly the same as the regular languages considered above. This result is important also for the characterization of the behavior of finite automata studied in Chapter 3.

Definition 2.5.   *A Post system, PS, is **regular** iff every production is of the form $uB \to wB$, where B is a variable and u and w are words over the alphabet of constants. A language is called **RPS** iff it is generated by some regular Post system. A language is termed **PRPS** (short for generated by a pure regular Post system) iff it is of the form $L(PS, \Sigma_C)$, where PS is a regular Post system and Σ_C is the alphabet of constants of PS.*   □

We now begin the proof of the fact that a language is RPS iff it is regular. The following lemma proves this implication in one direction.

Lemma 2.6.   *Every regular language is PRPS.*

Proof.   Let L be an arbitrary regular language over the alphabet Σ. Since clearly the empty language is *PRPS*, we assume $L \neq \varnothing$. By Theorem 2.1, we assume that L is generated by a grammar G, where all productions are of one of the forms $A \to aB$ or $A \to \lambda$ and, furthermore, the uniqueness condition is satisfied. Let n be the cardinality of the nonterminal alphabet of G. Let S' be the start letter of G.

We now construct a regular Post system, PS, as follows. The alphabet of variables of PS consists of the letter B alone. (In fact, this can always be assumed of a regular Post system, as well as of a Post system in the normal form of Theorem 2.5.) The alphabet of constants equals Σ. The set of axioms consists of all words in L with length less than or equal n. (Clearly, these words can effectively be determined by considering derivations according to G.) The set of productions of PS is defined as follows.

Consider derivations of length n according to G such that the last word is not over the terminal alphabet. They are of the form

$$S' \Rightarrow a_1 A_1 \Rightarrow a_1 a_2 A_2 \Rightarrow \cdots \Rightarrow a_1 \cdots a_n A_n, \tag{17}$$

where, as usual, the capital letters are nonterminals and the lowercase letters are terminals. It is, of course, possible that some of the terminals, as well as some of the nonterminals, in (17) are equal. In fact, the definition of n guarantees that at least two of the nonterminals in (17) are equal. We let $S' = A_0$ and (i, j) be the first pair such that $A_i = A_j$. More specifically, let $j \leq n$ be the smallest number such that the nonterminal A_j occurs already earlier in derivation (17), and let $i < j$ be the unique number such that $A_i = A_j$. The pair (i, j) is referred to as the **repetition pair** associated with (17).

Clearly, there are only finitely many derivations like (17). Each of them is uniquely determined by the word $a_1 \cdots a_n$. To each of them we associate the repetition pair (i, j) and the production

$$a_1 \cdots a_i B \rightarrow a_1 \cdots a_i \cdots a_j B. \tag{18}$$

(For $i = 0$, the left side of this production reduces to B.) The production set of PS consists of all productions obtained in this fashion.

This completes the definition of PS. Clearly, PS is a regular Post system.

We shall now prove that

$$L = L(PS, \Sigma), \tag{19}$$

which shows that the language L is $PRPS$.

We establish first the inclusion $L(PS, \Sigma) \subseteq L$. Since the axioms are in L by definition, it suffices to prove for an arbitrary production (18) and an arbitrary word w that, whenever $a_1 \cdots a_i w$ is in L, then so is $a_1 \cdots a_i \cdots a_j w$. Since (i, j) is a repetition pair, there is a derivation according to G,

$$S' \Rightarrow^* a_1 \cdots a_i A_i \Rightarrow^* a_1 \cdots a_i \cdots a_j A_j, \tag{20}$$

where $A_i = A_j$. Moreover, since $a_1 \cdots a_i w$ is in L, there is a derivation according to G,

$$S' \Rightarrow^* a_1 \cdots a_i A_i \Rightarrow^* a_1 \cdots a_i w. \tag{21}$$

(Observe that the uniqueness property of G, mentioned in the second sentence of Theorem 2.1, guarantees that the nonterminal A_i is uniquely determined by the prefix $a_1 \cdots a_i$. Consequently, we have the *same A_i* in both (20) and (21). Without the uniqueness property, this would not necessarily have to be the case, and our argument would be invalid.) Combining the derivations of (20) and (21), we obtain the derivation

$$S' \Rightarrow^* a_1 \cdots a_i A_i \Rightarrow^* a_1 \cdots a_i \cdots a_j A_i \Rightarrow^* a_1 \cdots a_i \cdots a_j w,$$

which shows that the word $a_1 \cdots a_i \cdots a_j w$ is in L.

To establish the reverse inclusion, $L \subseteq L(PS, \Sigma)$, we assume the contrary and let w be the shortest word in the difference $L - L(PS, \Sigma)$. (More specifically, w is in the difference, and there is no word in the difference shorter than w.) By the choice of the axiom set of PS, we have $|w| > n$. We write $w = a_1 \cdots a_n w'$, where the a's are letters. Since w is in L, there is a derivation

$$S' \Rightarrow^* a_1 \cdots a_n A_n \Rightarrow^* a_1 \cdots a_n w' \tag{22}$$

according to G. Let (i, j) be the repetition pair associated with the initial part of (22)—that is, to the part given explicitly in (17). Since $A_i = A_j$, the derivation

$$S' \Rightarrow^* a_1 \cdots a_i A_i \Rightarrow^* a_1 \cdots a_i a_{j+1} \cdots a_n A_n$$
$$\Rightarrow^* a_1 \cdots a_i a_{j+1} \cdots a_n w'$$

is also valid according to G, which shows that the word $w_1 = a_1 \cdots a_i a_{j+1} \cdots a_n w'$ is in L. On the other hand, w_1 is not in $L(PS, \Sigma)$ because otherwise an application of production (18) for $B = a_{j+1} \cdots a_n w'$ gives the impossible result that w is in $L(PS, \Sigma)$.

Consequently, w_1 is in the difference $L - L(PS, \Sigma)$. Since $i < j$, we have $|w_1| < |w|$, which contradicts the choice of w. Hence, the inclusion $L \subseteq L(PS, \Sigma)$ follows.   □

Lemma 2.6 shows that for every regular grammar G, there is a regular Post system, PS, such that PS, when viewed as a pure system, is equivalent to G in the sense that it generates the same language as G. Moreover, our proofs give an *effective procedure* for constructing PS, given G. Unless explicitly stated otherwise, this is true of all of our proofs: A proof for the existence of a certain object always gives an effective procedure for constructing the object. The effectiveness of constructions is, of course, quite essential in any theory of computing. This issue will be further discussed in later chapters.

We now begin the proof of the converse of Lemma 2.6: If a language is *PRPS*, then it is regular. When more facts about regular languages become available in the next chapter, this result immediately yields the stronger result that every *RPS* language is regular.

The converse of Lemma 2.6 is established by two successive reductions, stated as Lemmas 2.7 and 2.8. A third lemma, Lemma 2.9, then shows how the result of these reductions can be generated by a regular grammar.

We need some further notations and terminology. Consider a regular Post system, $PS = (\Sigma_V, \Sigma_C, P, S)$. Because of the form of the productions, we may assume without loss of generality that Σ_V consists of a single letter B. This does not affect any of the languages $L(PS, \Sigma_T)$, where $\Sigma_T \subseteq \Sigma_C$. We consider an arbitrary such Σ_T, and let $\Sigma_N = \Sigma_C - \Sigma_T$. (It is possible that Σ_N is empty.) The system PS is called **reduced** iff every production is of one of the forms

$$A_1 a B \rightarrow A_2 B, \qquad A_1 B \rightarrow A_2 B, \qquad A_1 B \rightarrow A_2 a B,$$

where a is in Σ_T and A_1, A_2 are in Σ_N. These three forms are referred to as **contractions**, **neutrations** and **expansions**, respectively.

We also introduce a binary relation $\Rightarrow$ (yields directly) on the set Σ_C^* in the natural way: $\alpha_1 \Rightarrow \alpha_2$ holds for words α_1 and α_2 in Σ_C^* iff there is a word β over Σ_C such that

$$\alpha_1 = u\beta \quad \text{and} \quad \alpha_2 = w\beta$$

for some production $uB \rightarrow wB$ in the production set P. (Thus, the relation $\Rightarrow$ is defined in terms of PS and could be denoted $\Rightarrow_{PS}$.) As usual, $\Rightarrow^*$ denotes the reflexive, transitive closure of $\Rightarrow$. Clearly, $L(PS, \Sigma_C) = \{w \in \Sigma_C^* \mid s \Rightarrow^* w, \text{ for some } s \in S\}$.

Lemma 2.7. *For every regular Post system* $PS = (\Sigma_V, \Sigma_C, P, S)$, *there is a reduced regular Post system,* $PS' = (\Sigma_V, \Sigma_C', P', S)$, *where* $\Sigma_C \subsetneq \Sigma_C'$, *an element* $A_1 \in \Sigma_C' - \Sigma_C$, *and a subset* $F \subseteq \Sigma_C' - \Sigma_C$ *such that*

$$L(PS, \Sigma_C) = \{w \in \Sigma_C^* \mid f \Rightarrow_{PS'}^* A_1 w, \text{ for some } f \text{ in } F\}. \qquad (23)$$

Proof. Consider a sequence $A_1, A_2, \ldots$ of letters not in Σ_C. Define

$$PS_1 = (\Sigma_V, \Sigma_C \cup \{A_1\}, P_1, S),$$

where P_1 consists of all productions $A_1 uB \rightarrow A_1 wB$ such that $uB \rightarrow wB$ belongs to P. Denote

$$S_1 = \{A_1 x \mid x \text{ in } S\}.$$

Then

$$L(PS, \Sigma_C) = \{w \in \Sigma_C^* \mid s_1 \Rightarrow_{PS_1}^* A_1 w, \text{ for some } s_1 \in S_1\}. \qquad (24)$$

Having defined the pair (PS_i, S_i), $i \geq 1$, we define the pair (PS_{i+1}, S_{i+1}) by one of Rules 1–3, as long as at least one of the rules is applicable. The following invariance will be observed in the transition from PS_i to PS_{i+1}, for any $i \geq 1$:

$$\{w \in \Sigma_C^* \mid s_i \Rightarrow_{PS_i}^* A_1 w, \text{ for some } s_i \text{ in } S_i\}$$
$$= \{w \in \Sigma_C^* \mid s_{i+1} \Rightarrow_{PS_{i+1}}^* A_1 w, \text{ for some } s_{i+1} \text{ in } S_{i+1}\}. \qquad (25)$$

Rule 1. Assume the set S_i contains a word $A_j aw$, where $j \geq 1$, a is in Σ_C, and w is in Σ_C^*. In this case PS_{i+1} is obtained from PS_i by adding the new letter A_v and the production $A_v B \rightarrow A_j aB$. The set S_{i+1} is obtained from S_i by removing $A_j aw$ and adding $A_v w$.

Rule 2. Assume that PS_i contains a production $A_j awB \rightarrow uB$ that is not a contraction, where $a \in \Sigma_C$. In this case $S_{i+1} = S_i$, and PS_{i+1} is obtained from PS_i by adding the new letter A_v and the productions $A_v wB \rightarrow uB$ and $A_j aB \rightarrow A_v B$, as well as by removing the production $A_j awB \rightarrow uB$.

Rule 3. Assume that PS_i contains a production $uB \rightarrow A_j awB$ that is not an expansion, where $a \in \Sigma_C$. In this case $S_{i+1} = S_i$, and PS_{i+1} is obtained from PS_i by adding the new letter A_v and the productions $uB \rightarrow A_v wB$ and $A_v B \rightarrow A_j aB$, as well as by removing the production $uB \rightarrow A_j awB$.

It can now be immediately verified that (25) holds true, no matter which of Rules 1–3 has been used in defining PS_{i+1} and S_{i+1}. The order of the application of the rules is by no means unique. However, after finitely many applications, a pair (PS_k, S_k) is reached such that none of the rules is applicable. (This is a consequence of the following observations. Rule 1 shortens the words in the S-sets, while Rules 2 and 3 make the other side of some production strictly shorter. By the definition of the production set P_1, both sides of all productions have a nonterminal letter as a prefix.) It follows that PS_k is reduced and S_k consists of some of the letters A_j. Consequently, by (24) and (25), (23) holds with

$$PS' = PS_k, \qquad F = S_k. \qquad \square$$

Lemma 2.7 shows how to represent an arbitrary $PRPS$ language in the form of (23), where PS' is reduced. A reduced regular Post system still may have both contractions and expansions, which is awkward from the point of view of regular grammars. Therefore, a further simplification, presented in the next lemma, is needed. After this simplification, we are then able (in Lemma 2.9) to establish our goal.

Lemma 2.8. *Consider the language $L(PS, \Sigma_C)$, as well as PS', F, and A_1 as defined in Lemma 2.7. There is a reduced regular Post system PS'' not containing any expansions such that*

$$L(PS, \Sigma_C) = \{w \in \Sigma_C^* \mid A_1 w \Rightarrow_{PS''}^* f, \text{for some } f \text{ in } F\}. \tag{26}$$

Proof. First consider the representation in (23). Transform the system PS' into a system $\overline{PS}'$ by "reversing" the productions: $uB \to wB$ is in $\overline{PS}'$ iff $wB \to uB$ is in PS'. Clearly, $\overline{PS}'$ is reduced: Contractions just become expansions, and vice versa. It is also obvious that (26) holds, with PS'' replaced by $\overline{PS}'$.

To establish the lemma, we now modify $\overline{PS}'$ into a system PS'' without expansions such that (26) holds. The alphabets of PS'' are those of $\overline{PS}'$. We denote the terminal and nonterminal alphabets by Σ_T and Σ_N. (Observe that Lemmas 2.7 and 2.8 do not use the axiom set at all. So we may assume that the new systems constructed have the same axiom set as PS.)

The productions of PS'' are defined as follows. For a in Σ_T and C_1, C_2 in Σ_N, the contraction $C_1 aB \to C_2 B$ is in PS'' iff $C_1 a \Rightarrow_{\overline{PS}'}^* C_2$. The neutration $C_1 B \to C_2 B$ is in PS'' iff $C_1 \Rightarrow_{\overline{PS}'}^* C_2$. This completes the definition of PS''. We still have to establish (26).

Since (26) holds with PS'' replaced by $\overline{PS}'$—that is,

$$L(PS, \Sigma_C) = \{w \in \Sigma_C^* \mid A_1 w \Rightarrow_{\overline{PS}'}^* f, \text{ for some } f \text{ in } F\} \tag{27}$$

—we conclude that the right side of (26) is included in the left side. Conversely, assume that w is a word belonging to the left side of (26). If $w = \lambda$, then $A_1 \Rightarrow_{\overline{PS}'}^* f$ holds for some f in F. But this means that the neutration $A_1 B \to fB$ is in PS'', which implies that w belongs to the right side of (26).

Thus, assume that $w \neq \lambda$ belongs to the left side of (26). Write w in the form $w = a_1 \cdots a_k$, where the a's are letters. Thus, in view of (27),

$$A_1 a_1 \cdots a_k \Rightarrow_{\overline{PS}'}^* f$$

for some f in F. This derivation is possible only if the corresponding productions are in PS'', by the definition of PS''. Hence, w belongs to the right side of (26), completing the proof. The reader might still want to show in detail that the construction of PS'' is effective.   □

Lemma 2.9.   *Every PRPS language is regular.*

Proof.   Given a language $L(PS, \Sigma_C)$, we first construct, by Lemma 2.8, a reduced regular Post system, PS'', not containing any expansions such that (26) holds. We then construct a regular grammar G such that

$$L(G) = L(PS, \Sigma_C). \tag{28}$$

The terminal and nonterminal alphabets of G equal those of PS''. (Thus, the terminal alphabet is Σ_C and the nonterminal alphabet is $\Sigma_C' - \Sigma_C$. In the proof of Lemma 2.8, the two alphabets were denoted also by Σ_T and Σ_N.) The start letter of G is A_1.

The production set of G consists of all productions obtained as follows. For every nonterminal A in the set F, the production $A \rightarrow \lambda$ is in G. For every contraction $C_1aB \rightarrow C_2B$ in PS'', the production $C_1 \rightarrow aC_2$ is in G. For every neutration $C_1B \rightarrow C_2B$ in PS'', the production $C_1 \rightarrow C_2$ is in G. Clearly, G is a regular grammar. Also the validity of (28) is immediate.   □

By Theorem 2.1, we may, of course, eliminate the productions $C_1 \rightarrow C_2$ from G. This also shows how a stronger version of Lemma 2.8, where PS'' contains neither expansions nor neutrations, can be obtained.

The following theorem is obtained by combining Lemmas 2.6 and 2.9. It will be seen in the next chapter that the statement holds true also with *PRPS* replaced by *RPS*.

Theorem 2.10.   *A language is PRPS iff it is regular.*

2.4.   MARKOV ALGORITHMS

The model studied in this section resembles the intuitive notion of an algorithm: It gives rise to an effective procedure for processing an input. At each step of the procedure, the next step is uniquely determined. This property of *monogenicity* is possessed neither by grammars nor by Post systems, where in the course of a derivation we may have several choices. However, viewed as language-defining devices, Markov algorithms are equivalent to grammars and Post systems.

Definition 2.6.   *A Markov algorithm is a rewriting system (Σ, P), where the elements of P are given in a linear order*

$$\alpha_1 \rightarrow \beta_1, \ldots, \alpha_k \rightarrow \beta_k, \tag{29}$$

and, furthermore, a subset P_1 *of P is given. The elements* (α, β) *of* P_1 *are called final productions and are denoted by* $\alpha \to . \beta$.   □

At each step of the rewriting process, the first applicable production from (29) must be chosen and, furthermore, the leftmost occurrence of its left side must be rewritten. The rewriting process terminates when some final production is applied or when none of the productions (29) is any more applicable. These notions are formalized in our next definition.

Definition 2.7. *Given a Markov algorithm, MA, with productions (29), a binary relation* $\Rightarrow_{MA}$—*or, briefly,* $\Rightarrow$ *(yields directly)—on the set of all words over the alphabet of MA is defined as follows. For any words u and w, u* $\Rightarrow$ *w holds iff each of the conditions (i)–(iv) is satisfied.*

(i) *There is a number i,* $1 \le i \le k$, *and words* u_1, u_2 *such that* $u = u_1 \alpha_i u_2$, $w = u_1 \beta_i u_2$.

(ii) *None of the words* α_j *with* $j < i$ *is a subword of u.*

(iii) α_i *occurs as a subword of* $u_1 \alpha_i$ *only once.*

(iv) *One of the words* $\alpha_1, \ldots, \alpha_k$ *is a subword of w, and* $\alpha_i \to \beta_i$ *is not a final production.*   □

We may, of course, consider the reflexive, transitive closure $\Rightarrow^*$ of the relation $\Rightarrow$ introduced above. However, we do this differently, implementing the applications of final productions.

Definition 2.8. *For any words u and w over the alphabet of MA, u* $\Rightarrow$ *.w holds iff conditions (i)–(iii) in Definition 2.7 are satisfied but condition (iv) is not satisfied (that is, either* $\alpha_i \to \beta_i$ *is a final production, or none of the words* $\alpha_1, \ldots, \alpha_k$ *is a subword of w). Furthermore, u* $\Rightarrow^*$ *w holds iff there is a finite sequence of words*

$$u = x_0, x_1, \ldots, x_n, w, \qquad n \ge 0$$

such that $x_j \Rightarrow x_{j+1}$ *holds for* $0 \le j \le n - 1$ *and* $x_n \Rightarrow . w$, *or else u = w and none of the words* $\alpha_1, \ldots, \alpha_k$ *is a subword of u.*   □

It is an immediate consequence of the definitions that for any word u, there is at most one word w with the property $u \Rightarrow^* w$. If such a w exists, we say that our Markov algorithm MA **halts** with u and **translates** u into w. Otherwise, we say that MA **loops** with u.

For instance, the Markov algorithm with the alphabet $\{a, b\}$ and the list of productions

$$a \to \lambda, \qquad b \to \lambda, \qquad \lambda \to . abba$$

translates every word over the alphabet $\{a, b\}$ into the word *abba*. The process is very simple: First all letters a are erased, beginning from the left, then all letters b are similarly erased, and, finally, the word *abba* is produced.

The Markov algorithm with the alphabet $\{a, b\}$ and the list of productions

$$a \to a^2, \qquad b^2 \to \lambda, \qquad b \to .ab$$

loops with every word containing an occurrence of a and translates, for every $i \geq 0$, the word b^{2i} (resp. b^{2i+1}) into λ (resp. ab).

The following example contains more-complicated Markov algorithms.

Example 2.7. The Markov algorithm with the alphabet $\{a, A, B, \#\}$ and the list of productions

$$Ba \to aB, \qquad Aa \to aBA, \qquad A \to \lambda, \qquad a\# \to \#A, \qquad \#a \to \#,$$
$$\# \to \lambda, \qquad B \to a$$

translates every word of the form

$$a^i \# a^j \quad (i, j \geq 0) \tag{30}$$

into the word a^{ij}. Hence, in this notation, this algorithm multiplies two numbers.

The Markov algorithm with the alphabet $\{a, A, B, C, \#\}$ and the list of productions

$$aA \to Aa, \qquad a\#a \to A\#, \qquad a\# \to \#B, \qquad B \to a, \qquad A \to C,$$
$$C \to a, \qquad \# \to \lambda$$

translates every word of (30) into the word a^k, where k is the greatest common divisor of i and j. (For $i = j = 0$, the result is 0.)

The verification of these facts is left to the reader. For both of these Markov algorithms, the set of final productions is empty. $\qquad \square$

So far we have considered Markov algorithms only as devices *translating* words or languages into words or languages. As we have already seen in connection with rewriting systems in general, as well as in connection with Post systems, such a translating device can be converted into a device defining a language. There are two different ways of doing this. We may either specify a (usually finite) set of start words and consider the collection of all words derivable from the start words, or else specify a set of end words and consider the collection of all words such that some end word is derivable from them. The former approach gives rise to a **generative** device for defining a language: We begin with the start words, and consider the words generated by them. The latter approach gives rise to an **accepting** device, or **acceptor,** for defining a language: Only words in the language lead to "acceptable" end words. Derivations according to a generative device *end* with words in the language, whereas those according to an accepting device *begin* with words in the language. Grammars are typical examples of generative devices, whereas various types of automata can be viewed as acceptors.

Both of these approaches can be augmented with the use of some auxiliary letters not appearing in the words of the language to be defined. The right sides of (26) and (27) provide examples of languages defined by an accepting device, where auxiliary letters are used and F is the set of end words. On the other hand, the right sides of (23) and (24) are examples of languages defined by a generative device, where again auxiliary letters are used and F (resp. S_1) is the set of start words.

It should be emphasized that in many cases the structure of a translating device makes it more natural to convert it into a generating device than into an accepting device, or vice versa. However, if the translating device is symmetric in regard to the form of the productions (such as a general rewriting system), then it can equally well be converted into a generating or into an accepting device.

How is it with Markov algorithms: Is it better to convert them into a generative or into an accepting device? Recall that Markov algorithms have the property that for any word u, there is at most one word w such that $u \Rightarrow^* w$. Consequently, a finite set of axioms is able to generate only a finite language. For this reason, it is natural to convert a Markov algorithm into an accepting device.

Definition 2.9.   Let MA be a Markov algorithm and Σ_T a subset of its alphabet. The language **accepted** by MA with respect to Σ_T is defined by

$$L(MA, \Sigma_T) = \{w \in \Sigma_T^* \mid w \Rightarrow^* \lambda\}. \tag{31}\ \square$$

For instance, the Markov algorithm MA with the list of productions

$$ab \to S, \qquad aSb \to S, \qquad S \to . \lambda$$

satisfies

$$L(MA, \{a, b\}) = \{a^i b^i \mid i \geq 0\}.$$

This equation is also satisfied by the Markov algorithm MA with the list of productions

$$aabb \to ab, \qquad ab \to . \lambda.$$

(Observe that it is essential that the second production is final. Otherwise, words having several occurrences of ab, such as $aabbab$, will also be accepted.)

The very simple Markov algorithm with

$$ab \to \lambda$$

as the only production accepts (with respect to $\{a, b\}$) the language $L(G_2)$ of properly nested parentheses, considered at the beginning of Section 2.2.

The reader should be warned: Grammars can not always be transformed into Markov algorithms just by reversing the productions (that is, interchanging the two sides of the productions) and adding the final production $S \to$. λ to the end of the list. The reason for this is that the productions of a Markov algorithm cannot be used arbitrarily as the productions of a grammar. For instance, consider the grammar

$$S \to \lambda, \qquad S \to SS, \qquad S \to aSb, \qquad S \to bSa,$$

generating the language considered in Example 2.4. The Markov algorithm, constructed as

$$\lambda \to S, \qquad SS \to S, \qquad aSb \to S, \qquad bSa \to S, \qquad S \to . \lambda,$$

does not define the same language. In fact, it defines the empty language, since it loops with every word.

However, by first modifying a given grammar to a form making a direct simulation by a Markov algorithm possible, one can prove the *if*-part of the following theorem. Information about the proof of the only *if* part is contained in Exercise 2.11.

Theorem 2.11. *A language is of the form in* (31), *for some Markov algorithm MA and alphabet* Σ_T, *iff it is recursively enumerable.*

2.5.   *L* SYSTEMS

All the models considered in Sections 2.1–2.4 have the property that at each step of the rewriting process, only some specific part of the word considered is actually rewritten. This is to be contrasted with *parallel* rewriting: At each step of the process, all letters of the word considered must be rewritten.

The most-widely studied parallel rewriting systems are referred to as *L* **systems;** they were introduced by Lindenmayer [Li] to provide a model for the development of filamentous organisms. In this connection parallel rewriting is most natural, since the development takes place simultaneously everywhere in the organism rather than in some specific part only. Of course, words as such represent one-dimensional organisms, but by suitable conventions one may use words to describe the many-dimensional case as well, [RozS].

To take a very simple example, consider the production $a \to aa$. When it is viewed as a production in a rewriting system, all words a^i, $i \geq 1$, are derivable from the start word a. On the other hand, if parallel rewriting is considered, then only words of the form a^{2^i}, $i \geq 0$, are derivable from the start word a. This follows because, for instance, the word a^2 directly yields only the word a^4 because the production $a \to aa$ has to be applied for *both* occurrences of the letter a. If we have the productions $a \to aa$ and $a \to a$ available, then all words a^i, $i \geq 1$, are derivable from the start word a, even if parallel rewriting is considered. Thus, the presence of the production $a \to a$, which is immaterial for rewriting systems, makes a big difference if parallel rewriting is considered: It makes possible for us to keep some occurrences of a unaltered while the derivation is carried on in some other parts of the word.

We now define the most-widely studied types of L systems, 0L and D0L systems. We use the notations customary in the theory of L systems: D stands for *deterministic* and 0 stands for *context-free,* the latter abbreviation being based on the original biological connotation: The communication between different cells in the developmental process is 0-sided—that is, there is no communication at all.

Definition 2.10.   *A* 0L *system is a triple,* $G = (\Sigma, P, w)$, *where* Σ *is an alphabet,* w *is a word over* Σ *(start word or axiom), and* P *is a finite set of ordered pairs* (a, α), *where* a *is in* Σ *and* α *is in* Σ^*, *such that for each* a *in* Σ, *there is at least one pair* (a, α) *in* P. *The elements* (a, α) *of* P *are called productions and are denoted by* $a \rightarrow \alpha$. *If, for each* a *in* Σ, *there is exactly one production* $a \rightarrow \alpha$ *in* P, *then* G *is called a* D0L *system.*   □

The above definition defines 0L systems as rewriting systems of a special kind. However, we do not want to call them rewriting systems because the binary relation ⇒ (yields directly) is defined differently for 0L systems and for rewriting systems. The definition for 0L systems, which takes care of the parallelism, will now be given.

Definition 2.11.   *Given a* 0L *system* $G = (\Sigma, P, w)$, *a binary relation* ⇒$_G$—*or, briefly,* ⇒ —*on the set* Σ^* *is defined as follows:*

$$a_1 \cdots a_n \Rightarrow \alpha_1 \cdots \alpha_n, \qquad n \geq 1, \text{ each } a_i \in \Sigma \tag{32}$$

holds iff $a_i \rightarrow \alpha_i$ *is in* P *for each* $i = 1, \ldots, n$. *Moreover,* $\lambda \Rightarrow \lambda$ *holds. The language generated by* G *is defined by*

$$L(G) = \{u \mid w \Rightarrow^* u\}, \tag{33}$$

where ⇒* *is the reflexive, transitive closure of* ⇒. *Languages of the form of* (33), *where* G *is a* 0L *(resp.* D0L*) system, are referred to as* 0L *(resp.* D0L*) languages.*   □

Observe that in (32) it is not required that the a_i's are distinct. Two occurrences of the same letter may be rewritten differently. For instance, if G has the productions $a \rightarrow \alpha$ and $a \rightarrow \beta$, then all the following relations hold:

$$aa \Rightarrow \alpha\alpha, \qquad aa \Rightarrow \alpha\beta, \qquad aa \Rightarrow \beta\alpha, \qquad aa \Rightarrow \beta\beta.$$

However, in case of a D0L system, for each word u over the alphabet of the system, there is exactly one word u' such that $u \Rightarrow u'$. (Because of our convention to the effect that $\lambda \Rightarrow \lambda$, this holds true also for $u = \lambda$.) Consequently, a *D0L* system *G* defines the following unique **sequence** *E*(*G*) of words over Σ:

$$w_0 = w, w_1, w_2, \ldots,$$

where $w_i \Rightarrow w_{i+1}$ holds for every $i \geq 0$. Such sequences *E*(*G*) are referred to as **D0L sequences**.

Clearly, the words in the sequence *E*(*G*) are exactly the same as the words in the language *L*(*G*). Thus, a *D0L* system is a generating device defining its language in a specific linear *order*. None of the other devices we have considered so far provides such an ordering of the language defined.

Since in a *D0L* system *G* there is exactly one production for each letter *a* of Σ, the system *G* defines a morphism $h : \Sigma^* \rightarrow \Sigma^*$ in the natural way: $h(a) = \alpha$ iff $a \rightarrow \alpha$ is a production in *G*. The sequence *E*(*G*) consists, in this notation, of the words

$$h^0(w) = w, h^1(w) = h(w), h^2(w) = h(h(w)), \ldots.$$

Thus, a *D0L* sequence is obtained by *iterating* the morphism *h*.

For instance, for the *D0L* system $G = (\{a\}, \{a \rightarrow a^2\}, a)$, we have $L(G) = \{a^{2^i} \mid i \geq 0\}$. The sequence *E*(*G*) consists of the words in *L*(*G*) ordered according to increasing length. This example shows clearly how parallelism opens a new dimension for the definition of languages. Any grammar generating the same language *L*(*G*) is fairly complicated, as the reader may verify.

The *0L* system with the axiom *a* and productions $a \rightarrow a$, $a \rightarrow a^2$ generates the language $\{a^i \mid i \geq 1\}$. This language is not a *D0L* language because every *D0L* system with the alphabet $\{a\}$ has only one production $a \rightarrow a^i$ for some $i \geq 0$.

In the sequel we consider again somewhat more complicated examples.

Example 2.8. Consider the *D0L* system $G = (\{a, b\}, \{a \rightarrow ab, b \rightarrow ba\}, a)$. The first few words in the sequence *E*(*G*) are

$$a, \quad ab, \quad abba, \quad abbabaab, \quad abbabaabbaababba. \quad (34)$$

In general, it is easy to verify inductively that, for every $i \geq 1$, the $(i + 1)$st word w_{i+1} in the sequence satisfies

$$w_{i+1} = w_i w_i',$$

where w_i' results from w_i by interchanging *a* and *b*. This also shows that every word in the sequence *E*(*G*) is a prefix of the next one. (This property is, in general, a consequence of the fact that the axiom is a prefix of the next word in *E*(*G*).) Consequently, *G* defines an "infinite word" e_G—that is, an infinite sequence of letters of the alphabet $\{a, b\}$ in a natural fashion. To find the jth letter of e_G, we just look at the jth letter in some w_i with $|w_i| \geq j$. Thus, the last word in (34) gives us the first 16 letters of e_G, the next word in the sequence $E(G)$ gives us the first 32 letters, and so on.

The sequence e_G is the celebrated **Thue sequence,** due to Thue [Th1]. It has the property of **cube-freeness:** It has no subsequence of the form x^3, $x \neq \lambda$. The reader is referred to [Sa5] for various applications of this fact. The proof is also left to the reader. It should be not too difficult, following the subsequent guidelines: (1) Neither a^3 nor b^3 occurs as a subword in e_G because e_G is obtained by catenating the two words ab and ba. (2) Consequently, neither $ababa$ nor $babab$ occurs as a subword in e_G. (3) Whenever a^2 or b^2 occurs as a subword in e_G, starting with the jth letter of e_G, then j is even. By (2) and (3), it is now easy to establish inductively the cube-freeness of e_G, recalling the fact that e_G is obtained by iterating a morphism h.

It is clear that no word of length greater than or equal to 4 over a two-letter alphabet $\{a, b\}$ is square-free in the sense that it has no subword x^2, $x \neq \lambda$; beginning with a, the next letter has to be b and the following a, after which a square is unavoidable. On the other hand, it is possible to construct an infinite square-free word over a three-letter alphabet, using e_G. See Exercise 2.12.   □

As in connection with the other devices we have considered, auxiliary letters can be introduced for L systems as well. In this fashion their generative capacity is increased. The terminal alphabet provides an additional filtering mechanism: Words generated by the system have to pass through this filter in order to qualify to be included in the language. In the next definition we again use an abbreviation customary in the theory of L systems: the letter E means *extended*.

Definition 2.12. *An EOL system is a pair $G_1 = (G, \Sigma_T)$, where $G = (\Sigma, P, w)$ is a 0L system and Σ_T is a subset of Σ, referred to as the **terminal alphabet**. The language generated by G_1 is defined by*

$$L(G_1) = L(G) \cap \Sigma_T^*.$$

*Languages of this form are called **EOL languages.***   □

Given a context-free grammar G, for each letter α (both terminal and nonterminal) we add the production $\alpha \rightarrow \alpha$ and view the resulting construct G_1 as an EOL system. It is then clear that $L(G) = L(G_1)$. This shows that every context-free language is also an EOL language. The next example demonstrates how to generate the language studied in Example 2.5 by an EOL system. It will be seen in Chapter 3 that this language is not context-free.

Example 2.9. Consider the EOL system G with the axiom S and productions $S \rightarrow ABC$, $A \rightarrow A\bar{A}$, $B \rightarrow B\bar{B}$, $C \rightarrow C\bar{C}$, $\bar{A} \rightarrow \bar{A}$, $\bar{B} \rightarrow \bar{B}$, $\bar{C} \rightarrow \bar{C}$, $\bar{A} \rightarrow a$, $\bar{B} \rightarrow b$,

$\bar{C} \rightarrow c$, $A \rightarrow a$, $B \rightarrow b$, $C \rightarrow c$, $a \rightarrow N$, $b \rightarrow N$, $c \rightarrow N$, and $N \rightarrow N$, where $\{a, b, c\}$ is the terminal alphabet. It is not difficult to verify that

$$L(G) = \{a^n b^n c^n \mid n \geq 1\}.$$

The system G is based on the **synchronization** of terminals: the terminals have to be introduced simultaneously in the whole word. If terminals are introduced only in some parts of the word, then the derivation cannot be continued to yield a word over the terminal alphabet. This is due to the fact that terminals give rise only to the "garbage" letter N, which cannot be eliminated once it is introduced. $\square$

The *L* systems considered so far are based on context-free productions: Each letter is rewritten independently of its neighbors. This is indicated by the number 0 in the name of the system. We do not define formally the notion of an *IL* **system** (*I* meaning *interactions*). The productions of an *IL* system have the form

$$(\beta_1, a, \beta_2) \rightarrow \alpha, \tag{35}$$

meaning that the letter a can be rewritten as the word α *between* the words β_1 and β_2. Productions (35) are again applied *in parallel:* Each letter has to be rewritten during one step of the derivation process. This means that there must be at least one production of the form of (35) for each configuration (β_1, a, β_2), including the cases where a is near the beginning or the end of the word. It should be easy for the reader to come up with the formal details. Again, an *EIL* **system** means an *IL* system provided with a specific terminal alphabet. Languages generated by *IL* and *EIL* systems are referred to as *IL* and *EIL* **languages,** respectively.

As the reader might expect, *EIL* systems are general enough to generate all recursively enumerable languages. Some hints for the proof of the following theorem are contained in Exercise 2.13.

Theorem 2.12.   *A language is recursively enumerable iff it is an EIL language.*

An area quite essential in the study of *L* systems is the theory of *growth functions.* From the biological point of view, the growth function measures the size of the organism modeled. From the theoretical point of view, the theory of growth functions reflects a fundamental aspect of *L* systems: generating languages as *sequences* of words.

Definition 2.13.   *For a D0L system G, we associate the function f_G mapping the set of nonnegative integers into itself as follows. For $i \geq 0$, $f_G(i)$ is defined to be the length of the ith word in the sequences $E(G)$, whereby the axiom of G is understood to be the 0th word. The function f_G is called the* **growth function** *of G.* $\square$

For instance, for the $D0L$ system G with the axiom a and the only production $a \to aa$, we have $f_G(i) = 2^i$ for all $i \geq 0$.

For the $D0L$ system G with the axiom a and the productions $a \to ab$ and $b \to b$, we have $f_G(i) = i + 1$ for all $i \geq 0$.

For the $D0L$ system G with the axiom a and productions $a \to abc^2$, $b \to bc^2$, and $c \to c$ (resp. $a \to abd^6$, $b \to bcd^{11}$, $c \to cd^6$, and $d \to d$) we have $f_G(i) = (i + 1)^2$ (resp. $f_G(i) = (i + 1)^3$).

The **analysis problem** for growth functions consists of determining the growth function of a given $D0L$ system. The **synthesis problem** consists of materializing given functions—if possible—as $D0L$ growth functions. (Since we do not want to enter a detailed discussion at this point, we do not specify explicitly what we mean by "given." Basically, the function can be given by some "reasonable" formula.) Both problems can be solved by mathematical methods based on our next theorem.

To a given $D0L$ system G with the alphabet $\{a_1, \ldots, a_n\}$, we associate the **axiom vector** $\pi_G = (v_1, \ldots, v_n)$, where v_i equals the number of occurrences of a_i in the axiom of G for each $i = 1, \ldots, n$. We also associate with G the $n \times n$ **growth matrix** M_G whose (i, j)th entry equals the number of occurrences of a_j in the production for a_i for all i and j. Finally, η is a column vector of dimension n consisting entirely of 1s.

The next theorem about growth functions is easy to establish by an induction on the length of the words.

Theorem 2.13.   The growth function f_G associated with a $D0L$ system G satisfies

$$f_G(i) = \pi_G M_G^i \eta \quad \text{for all } i \geq 0.$$

It is a consequence of Theorem 2.13 that a growth function of a $D0L$ system cannot grow faster than exponentially, and if it is ultimately growing, it cannot grow slower than linearly. Our final example shows that the growth functions of IL systems, analogously defined, do not necessarily satisfy these conditions.

Example 2.10.   Consider the following *deterministic IL* system G with the alphabet $\{a, b, c, d\}$ and axiom ad. The productions of G are defined by the following table, where the row indicates the left context and the column the symbol to be rewritten. If the left context of a letter is λ, it is understood to mean that the letter has no left neighbors.

	a	b	c	d
λ	c	b	a	d
a	a	b	a	d
b	a	b	a	d
c	b	c	a	ad
d	a	b	a	d

Thus, for instance, an occurrence of a without a left neighbor (that is, a is the first letter in a word) must be rewritten as c, whereas an occurrence of d with the left neighbor c must be rewritten as ad. The first few words in the sequence determined by G are:

$$ad, cd, aad, cad, abd, cbd, acd, caad, abad, cbad, acad, cabd, abbd, cbbd,$$
$$acbd, cacd, abaad.$$

This length sequence grows very slowly. In fact, the lengths of the intervals in which the growth function stays constant grow exponentially. This means that the overall growth is only logarithmic. The system G is known as *Gabor's sloth* due to Gabor Herman [RozS].   □

EXERCISES

2.1.   Which of the following equations are valid for all languages L_1, L_2, and L_3 and all morphisms h? Give reasons in each case.

$$L_1(L_2 \cap L_3) = L_1L_2 \cap L_1L_3, \qquad (L_1^*)^* \qquad = L_1^*,$$
$$(L_1 \cup L_2)^* = L_1^* \cup L_2^*, \qquad (L_1L_2)^*L_1 \qquad = L_1(L_2L_1)^*,$$
$$(L_1 \cup L_2)^* = L_2^*(L_1L_2^*)^*, \qquad h(h(L_1)) \qquad = h(L_1),$$
$$h(L_1 \cup L_2) = h(L_1) \cup h(L_2), \qquad h(L_1 \cap L_2) = h(L_1) \cap h(L_2),$$
$$h(L_1L_2) \quad = h(L_1)h(L_2), \qquad h(L_1^*) \qquad = (h(L_1))^*.$$

2.2.   A context-free grammar is called **reduced** iff for every nonterminal A different from the start letter S, (1) S generates a word containing A and (2) A generates some terminal word. Give a method that produces, for every context-free grammar, an equivalent reduced grammar.

2.3.   Modify Theorem 2.2 in such a way that also the languages containing the empty word are taken care of. (Add a new start letter S_1 and the production $S_1 \to \lambda$ whenever needed.)

2.4.   Characterize the three languages obtained by considering only two of Rules 1–3 in Example 2.4. Conclude that Rules 1–3 are independent in the sense that none of them can be omitted without altering the language.

2.5.   Show that every recursively enumerable language is generated by a grammar in which every production is of one of the three forms $A \to BC$, $AB \to \lambda$, or $A \to a$. (Consult [Sav].)

2.6.   A grammar is called **context-sensitive** iff the right side of every production is at least of the same length as the left side. Languages generated by such grammars are also referred to as context-sensitive. (An exceptional production can be added to the grammar in order to generate the empty word.) Establish a normal-form result analogous to Theorem 2.3 for context-sensitive grammars. (Context-sensitive grammars and languages constitute one of the main classes in the original Chomsky hierarchy. However, this class has turned out to be rather insignificant from both theoretical and practical point of view.)

2.7.   Show that pure Post systems do not generate all recursively enumerable languages. (*Hint:* For instance, in the language $\{a^{2^n} \mid n \geq 1\}$, the gaps grow too large.)

2.8.   Show that the construction of the Post system PS'' in Lemma 2.8 is effective.

2.9.   Let PS be a regular Post system and let U and V be two finite sets of words over the alphabet of constants of PS. Show that the languages

$$L_g(PS, U, V) = \{w \in \Sigma_T^* \mid w_u \Rightarrow^* w_v w \text{ for some } w_u \in U, w_v \in V\}$$

and

$$L_a(PS, U, V) = \{w \in \Sigma_T^* \mid w_u w \Rightarrow^* w_v \text{ for some } w_u \in U, w_v \in V\}$$

are regular. (Intuitively, L_g is the language generated by the triple (PS, U, V) and L_a is the language accepted by the same triple.) Can the result be extended to concern the case where U and V are infinite regular languages?

2.10.   Consider the Markov algorithm MA whose alphabet consists of the terminal alphabet Σ_T and of three additional letters A, B, and C. The list of productions is as follows:

$$\alpha\beta A \rightarrow \beta A\alpha, \quad B\alpha \rightarrow \alpha A\alpha B, \quad A \rightarrow C, \quad C \rightarrow \lambda,$$
$$B \rightarrow . \lambda, \quad \lambda \rightarrow B,$$

where α and β independently range over Σ_T. (The mutual order of productions in the first two schemata is immaterial.) Prove that MA translates every word w over Σ_T into ww.

2.11.   Prove Theorem 2.11. (Observe that Markov algorithms can be simulated by grammars if *messengers* are used: After each rewriting step, the messengers check that the rewriting was done according to the rules of the Markov algorithm. If not, then the messengers introduce some garbage symbol that guarantees that the derivation will not terminate.)

2.12.   Construct an infinite square-free word over a three-letter alphabet. (*Hint:* Start with the infinite word of Example 2.8. Replace every occurrence of the letter a by a_a or a_b, depending on whether the following letter is a or b. Similarly, replace every occurrence of b by b_a or b_b.)

2.13.   Prove Theorem 2.12. (The simulation of a grammar, in a suitable normal form, by an *EIL* system is easy. The converse simulation is possible by the use of messengers to take care of the parallelism. On the other hand, the converse simulation becomes unnecessary if this exercise is postponed until the first two sections of Chapter 4 have been read.)

2.14. Estimate the growth function of the system G discussed in Example 2.10.

2.15. Construct a *DOL* system whose growth function equals $(i + 1)^4$.

CHAPTER 3

Restricted Automata

3.1 FINITE AUTOMATA

A finite automaton is a strictly finitary model of computation. Everything involved is of a fixed, finite size and cannot be extended during the course of computation. The other types of automata studied later have at least a potentially infinite memory. Differences between various types of automata are based mainly on how information can be accessed in the memory.

A finite automaton operates in discrete time, as do all essential models of computation. Thus, we may speak of the "next" time instant when specifying the functioning of a finite automaton.

The simplest case is the memoryless device, where, at each time instant, the output depends only on the current input. Such devices are models of combinational circuits.

In general, however, the output produced by a finite automaton depends on the current input as well as on earlier inputs. Thus, the automaton is capable (to a certain extent) of remembering its past inputs. More specifically, this means the following.

The automaton has a finite number of **internal memory states.** At each time instant i it is in one of these states, say q_i. The state q_{i+1} at the next time instant is determined by q_i and by the input a_i given at time instant i. The output at time instant i is determined by the state q_i (or by q_i and a_i together).

For instance, consider an automaton with three states, q_0, q_1, and q_2. Assume that there is only one possible input, denoted by a, which permutes the states cyclically. (Thus, the automaton goes to the state q_1 if it receives the input a in q_0, and so forth.) For $i = 0, 1, 2$, the automaton produces the output i in the state q_i. This automaton is a time counter modulo 3: At each moment the output indicates the residue class modulo 3 of the length of the whole input so far, provided the automaton is started in q_0. Thus, the automaton is able to count only modulo the number of its states. This phenomenon is very typical for finite automata in general.

A more sophisticated example is provided by an elevator in a building with a certain number of floors. The input consists of all requests made at any given moment. The state is determined by information about the still-unfulfilled requests (this includes information about which floors were requested from inside the elevator), about the present location of the elevator, as well as about the present direction (up or down). The total number of possible states is finite and depends on the number of floors. The next state is always determined by the present state and input. The present state also determines the output—that is, the behavior of the elevator.

Consider still another example, a lexical scanner.

Example 3.1. Suppose we want to find out whether or not one of the key words *WATER* and *BLOOD* appears in a given text. When reading the text one letter at a time, we have to remember at most four previous letters. This can be accomplished by the following finite automaton.

The automaton has ten states. We use for them the intuitive notations

$$X, W, WA, WAT, WATE, B, BL, BLO, BLOO, ACC.$$

Being in the state *WAT* means that the last three letters read constitute the subword *WAT*. An analogous situation holds in each of the other states denoted by a word beginning with *W* or *B*. Being in the state *X* means that the

TABLE 3.1

	W	A	T	E	R	B	L	O	D
X	W					B			
W	W	WA				B			
WA	W		WAT			B			
WAT	W			WATE		B			
WATE	W				ACC	B			
B	W					B	BL		
BL	W					B		BLO	
BLO	W					B		BLOO	
BLOO	W					B			ACC
ACC	ACC	ACC	ACC	ACC	ACC	ACC	ACC	ACC	ACC

word read so far does not contribute a single character towards a key word. Thus, initially, the automaton is in the state X. If the automaton is in the state ACC (*accept*), then an occurrence of a key word has been found.

The state transitions are given in Table 3.1. The present state is read from the rows, and the present input letter from the columns. The letter X appears in all empty spaces, including the spaces below the input letters not listed, except spaces in the row for ACC. Observe that this automaton checks only whether either $WATER$ or $BLOOD$ appears as a subword in the input text, when the latter is viewed as one word. Thus, both of the input texts $WATERLOO$ and $OTTAWATERMINAL$ lead to acceptance. To avoid this, we can provide the input with boundary markers, #, indicating the space between words, and check the appearance of $\#WATER\#$ or $\#BLOOD\#$ as a subword of the input thus modified.   □

Our discussion so far has been very informal. We have tried to emphasize the essential component in a finite automaton, the **transition table**, which gives the next state in terms of the present state and input. The table (which, in fact, defines the transition function) essentially specifies the behavior of the automaton. Depending on whether the automaton is viewed as a translating or language-defining device, some output mechanism or accepting mechanism has to be defined. (As a language-defining device, a finite automaton is naturally viewed as an acceptor.)

All these notions are formalized below. In the first two sections of this chapter, we consider finite automata as acceptors; the translator aspect is handled in Section 3.3. The most convenient starting point for the formal definitions is the notion of a rewriting system. It provides a uniform way to approach both accepting and generating devices.

Definition 3.1. *A rewriting system* (Σ, P) *is called a* *finite deterministic automaton iff conditions* (*i*)–(*iii*) *are satisfied*:

(i) Σ *is divided into disjoint alphabets* Σ_Q *and* Σ_T, *referred to as the* **state** *and the* **terminal** (*or the* **input**) *alphabet*.

(ii)   *An element* $q_0 \in \Sigma_Q$ *and a subset* $Q_F \subseteq \Sigma_Q$ *are specified, called the* **initial state** *and the set of* **final states**, *respectively*.

(iii)   *The productions in* P *are of the form*

$$q_i a \to q_j, \qquad q_i, q_j \in \Sigma_Q; \quad a \in \Sigma_T. \tag{1}$$

Furthermore, for each pair (q_i, a) *where* $q_i \in \Sigma_Q$ *and* $a \in \Sigma_T$, *there is exactly one production* (1) *in the set* P.

Let FDA be a finite deterministic automaton and let $\Rightarrow$ *be the direct-yield relation of the underlying rewriting system. Then the language* **accepted** *by FDA is defined by*

$$L(FDA) = \{w \in \Sigma_T{}^* \mid q_0 w \Rightarrow^* q_F \text{ for some } q_F \in Q_F\}.$$

A language L is acceptable by a finite deterministic automaton, in short, AFDA, iff L = L(FDA) for some finite deterministic automaton FDA. □

The time counter modulo 3 considered above is determined by the productions

$$q_0 a \to q_1, \qquad q_1 a \to q_2, \qquad q_2 a \to q_0.$$

Here $\Sigma_Q = \{q_0, q_1, q_2\}$ and $\Sigma_T = \{a\}$. The state q_0 is the initial one. If we agree that it is also the only final state, then the accepted language is

$$\{a^{3i} \mid i \geq 0\} = \{a^3\}^*.$$

Because of condition (iii) in Definition 3.1, the productions can also be specified in terms of a transition table that can be used to specify, at the same time, both the state and the terminal alphabet. In this case, the transition table looks like this:

	a
q_0	q_1
q_1	q_2
q_2	q_0

For Example 3.1, the transition table was already given.

The transition table can always be expressed as a **transition diagram.** This means a graph where the (labeled) nodes are the states and the (labeled) arrows indicate the state transitions. In case of our time counter modulo 3, the transition diagram is shown in Figure 3.1. The reader might want to draw the transition diagram for the Example 3.1.

Example 3.2. Consider the finite deterministic automaton FDA with

$$\Sigma_Q = \{q_0, q_1, q_2, q_3\}, \qquad \Sigma_T = \{a, b\}, \qquad Q_F = \{q_0, q_1\},$$

with initial state q_0 and productions defined as follows:

$$q_0 a \to q_2, \qquad q_0 b \to q_3, \qquad q_1 a \to q_3, \qquad q_1 b \to q_0,$$
$$q_2 a \to q_0, \qquad q_2 b \to q_3, \qquad q_3 a \to q_1, \qquad q_3 b \to q_0.$$

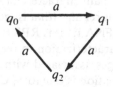

FIGURE 3.1

To characterize the language $L(FDA)$, we observe first that FDA is always in one of the states q_0 and q_2 (resp. q_1 and q_3) after receiving an even number greater than or equal to 0 (resp. an odd number greater than or equal to 1) of b's. Moreover, immediately after receiving the last occurrence of b, the automaton FDA is either in the state q_0 or q_3. Hence, if the input word contains an even (resp. odd) number of b's, then an even (resp. odd) number of a's is needed after the last occurrence of b to bring FDA to a final state. The occurrences of a before the last occurrence of b are immaterial. This characterizes the language $L(FDA)$.   □

Our next definition introduces a nondeterministic version of a finite automaton. In other words, the state transitions are not uniquely determined by the present state q and the input a. This means that for pairs (q, a), there may be several (including none at all) next states. And after reading a word w, the automaton can be in several states.

Since we consider several types of nondeterministic automata in the sequel, it should be emphasized that, as such, they are not intended to be feasible models for computation. However, they are very useful *auxiliary* concepts. Often, a nondeterministic automaton is suitable for a certain task. Then it is natural first to associate a nondeterministic device with this task and then to use results interconnecting nondeterministic devices of the type considered with the corresponding deterministic devices.

Definition 3.2.   A finite nondeterministic automaton FNA is defined as a deterministic one, with the following two exceptions. In (ii), q_0 is replaced by a subset $Q_I \subseteq \Sigma_Q$, referred to as the set of initial states. In (iii), the second sentence, "Furthermore,...," is omitted.
The language accepted by FNA is defined by

$$L(FNA) = \{w \in \Sigma_T^* \mid q_I w \Rightarrow^* q_F \text{ for some } q_I \in Q_I \text{ and } q_F \in Q_F\}.$$

A language L is acceptable by a finite nondeterministic automaton, in short AFNA, iff $L = L(FNA)$ for some finite nondeterministic automaton FNA.
                                                                    □

3.2.   KLEENE CHARACTERIZATION

In this section, we deduce the main characterization results for finite automata. We show that a language is regular (that is, generated by a regular grammar) iff it is AFDA iff it is AFNA iff it is RPS (in the sense of Section 2.3). Moreover, a very natural characterization in terms of simple operations is given for this class of languages associated with strictly finitary computing devices. The latter characterization is due to S. C. Kleene [Kl] and is, therefore, generally known as the **Kleene characterization.**

The three operations of union, catenation, and catenation closure (that is, star, *) form the basis of the Kleene characterization. We consider all languages obtainable from **atomic languages,** that is the empty language, $\varnothing$, and the singleton languages, $\{a\}$, where a is a letter of the alphabet, by finitely many applications of the three operations mentioned. The formula that tells us how a specific language is obtained in this fashion is referred to as a *regular expression.*

These notions are made explicit in our next definition.

Definition 3.3. *Assume that* Σ *and* $\Sigma' = \{\cup, *, \varnothing, (,)\}$ *are disjoint alphabets. A word* w *over the alphabet* $\Sigma \cup \Sigma'$ *is a regular expression over* Σ *iff* w *is a letter of* Σ *or the letter* $\varnothing$ *or else* w *is one of the forms* $(w_1 \cup w_2)$, $(w_1 w_2)$, *or* w_1^*, *where* w_1 *and* w_2 *are regular expressions over* Σ.

Each regular expression w *over* Σ *denotes a language* $L(w)$ *over* Σ *according to the following conventions:*

  (i) *The language denoted by* $\varnothing$ *is the empty language.*
  (ii) *The language denoted by* $a \in \Sigma$ *consists of the word* a.
  (iii) *For all regular expressions* w *and* u *over* Σ, *we have:*

$$L((w \cup u)) = L(w) \cup L(u), \quad L((wu)) = L(w)L(u),$$
$$L(w^*) = (L(w))^*.$$

For a language L, *we use the abbreviation DRE to mean that* L *is denoted by some regular expression.* $\qquad\qquad\square$

For instance, $(aa)^*$ is a regular expression over the alphabet $\Sigma = \{a\}$. It denotes the language over this alphabet consisting of all words of an even length. The same language is denoted by infinitely many regular expressions—for instance,

$$(aa)^* \cup (\varnothing a), \qquad (aa \cup aaaaaa)^*, \qquad ((aa)^*)^* (aa)^*.$$

We have omitted here, as we shall also omit in the sequel, some unnecessary parentheses. The order of strength of the three operations involved in a regular expression is the natural one: Catenation is performed before union and catenation closure before both union and catenation.

Thus, our purpose is to show that a language is regular iff it is AFDA iff it is AFNA iff it is RPS iff it is DRE. Many parts of this proof were carried out in Chapter 2. We still need the following two lemmas. The first one establishes a **closure property** for the family of languages generated by regular grammars: If certain languages are regular, then so are certain other languages obtained from the given ones by certain operations. Such closure properties have been widely studied in language theory (see Exercises 3.1–3.2). The proof of the second lemma analyzes the cycle structure in the transition diagram of a finite automaton.

Lemma 3.1.   *The union and catenation of two regular languages is regular and so is the catenation closure of a regular language.*

Proof.   Let $L_i = L(G_i)$, where G_i is a regular grammar for $i = 1, 2$. Since the nonterminals can always be renamed without affecting the generated language, we may further assume that the nonterminal alphabets of G_1 and G_2 are disjoint. Assume that S_1 and S_2 are the start letters of G_1 and G_2, respectively.

A regular grammar G_3 generating the union $L_1 \cup L_2$ is constructed as follows. The nonterminal alphabet of G_3 is the union of the nonterminal alphabets of G_1 and G_2, added to a new letter S, which is also the start letter of G_3. The terminal alphabet of G_3 is the union of the terminal alphabets of G_1 and G_2. The production set is the union of the production sets of G_1 and G_2, added to the two productions $S \rightarrow S_1$ and $S \rightarrow S_2$. (In fact, this same construction works for arbitrary and context-free grammars as well, showing that the families of recursively enumerable and context-free languages are closed under union.)

Of course, we can construct a grammar generating the catenation $L_1 L_2$ in exactly the same way, by just adding the production $S \rightarrow S_1 S_2$ instead of the two productions $S \rightarrow S_1$ and $S \rightarrow S_2$. The only problem with this approach is that the additional production $S \rightarrow S_1 S_2$ is not suitable for a regular grammar. (However, this approach is quite in order for context-free grammars, showing that the family of context-free languages is closed under catenation. The same approach works also for recursively enumerable languages; see Exercise 3.2.) Therefore, we have to modify this construction as follows.

We construct a regular grammar G_4 generating the catenation $L_1 L_2$ as follows. The nonterminal and terminal alphabet of G_4 is the union of the corresponding items of G_1 and G_2. The start letter of G_4 is S_1. The production set of G_4 consists of all productions of G_2 and, in addition, of all productions of G_1 modified in such a way that the terminating productions $A \rightarrow w$ (where w is a word over the terminal alphabet) are replaced by the productions $A \rightarrow w S_2$. It is now clear that $L(G_4) = L_1 L_2$. In fact, we always first derive an arbitrary word in L_1. But, instead of terminating the whole derivation, we deposit the nonterminal letter S_2 at the end of the word. After that, the field is open for derivations according to G_2.

Finally, a regular grammar G_5 generating the language $L_1{}^*$ is constructed as follows: Add to the nonterminal alphabet of G_1 a new start letter S, keep the terminal alphabet of G_1 fixed, and to the production set of G_1, add the productions $S \rightarrow \lambda$ and $S \rightarrow S_1$, as well as all productions obtained from the terminating productions $A \rightarrow w$ of G_1 by replacing them with $A \rightarrow w S_1$. (Thus, the original productions $A \rightarrow w$ are also kept in G_5.) It is then clear that $L(G_5) = L_1{}^*$. $\square$

Lemma 3.2. *Every AFDA language is DRE.*

Proof. Consider an arbitrary language $L = L(\text{FDA})$. Let Σ_T and $\Sigma_Q = \{q_1, \ldots, q_n\}$ be the terminal and state alphabet of FDA, where q_1 is the initial state. If the set Q_F of final states is empty, then L is denoted by the regular expression $\varnothing$. If Q_F consists of more than one states, then $L(\text{FDA})$ is the union of the languages obtained by replacing Q_F by each of the singleton sets $\{q\}$ with q in Q_F. Hence, without loss of generality, we assume that Q_F consists of exactly one element, say q_u.

Denote by

$$L_{ij}^k, \qquad 0 \le k \le n; \quad 1 \le i, j \le n,$$

the language consisting of all words w for which there is a derivation

$$q_i w \Rightarrow q_{i_1} w_1 \Rightarrow \cdots \Rightarrow q_{i_t} w_t \Rightarrow q_j,$$

where $i_v \le k$ for $v = 1, \ldots, t$. By definition, $\lambda \in L_{ij}^k$ iff $i = j$. (Intuitively, L_{ij} consists of all words causing a state transition in the automaton FDA from q_i to q_j in such a way that no state with an index greater than k appears as an intermediate state.) $\qquad \square$

Thus, for $i \ne j$, L_{ij}^0 is either empty or consists of some letters of Σ_T. For $i = j$, L_{ij}^0 consists of λ and of zero or more letters of Σ_T. Since λ is denoted by the regular expression $\varnothing^*$, we conclude that, for each i and j, L_{ij}^0 is denoted by a regular expression.

Proceeding inductively, we assume that, for a fixed value k with $0 \le k \le n - 1$, each of the languages L_{ij}^k is denoted by a regular expression. But clearly

$$L_{ij}^{k+1} = L_{ij}^k \cup L_{i(k+1)}^k \, (L_{(k+1)(k+1)}^k)^* \, L_{(k+1)j}^k \tag{2}$$

for all i, j, k. Indeed, consider an arbitrary word w taking FDA from q_i to q_j without using intermediate states with an index greater than $k + 1$. Then there are two possibilities. The word w may cause a state transition, where the state q_{k+1} does not appear at all as an intermediate state. In this case w belongs to the first term of the union on the right side of (2). The second possibility is that the state q_{k+1} actually appears as an intermediate state in the state transition caused by w. This means that w first causes a transition from q_i to q_{k+1}, then some loops from q_{k+1} to q_{k+1}, and, finally, a transition from q_{k+1} to q_j in such a way that q_{k+1} itself never appears as an intermediate state. But this implies that w belongs to the second term of the union on the right side of (2). Consequently, each of the languages L_{ij}^{k+1} is denoted by a regular expression. This completes the induction, showing that all our languages L_{ij}^k are DRE. Thus, the language $L = L(\text{FDA}) = L_{1u}^n$ is also DRE. $\qquad \square$

We are now in a position to establish the equivalences dealing with regular languages, to which we referred above. We want to show first that the following four properties are possessed simultaneously by every language L: (i) L is regular, (ii) L is AFDA, (iii) L is AFNA, and (iv) L is DRE.

Indeed, there are no more details to be discovered; everything is contained in our results established so far. We now summarize these results.

By Lemma 3.2, every AFDA language is DRE. By Lemma 3.1, every DRE language is regular. (More specifically, this follows because the atomic languages are clearly regular, and Lemma 3.1 shows that the family of regular languages is closed under the three operations involved.) Theorem 2.1 shows that every regular languauge is AFDA. (Indeed, the grammar obtained in Theorem 2.1 can be viewed as a finite deterministic automaton. The nonterminals are states, the start letter being the initial state. The set of final states consists of all states A for which the production $A \rightarrow \lambda$ is present. Each production $A \rightarrow aB$ is viewed as $Aa \rightarrow B$ in the rewriting system defining the automaton.) By definition, every AFDA language is an AFNA language, since a finite deterministic automaton is a special case of a finite nondeterministic one. Also by definition, every AFNA language is regular. (As above, we can convert the productions of a finite nondeterministic automaton into an equivalent regular grammar. The case where the FNA has several initial states is easily resolved.)

The interconnection of the arguments can be depicted by Figure 3.2.

The arguments used to establish this diagram show that the four devices involved (regular grammars, finite deterministic automata, finite nondeterministic automata, and regular expressions) are **equivalent** in the sense that they define the same family of languages and, moreover, that they are **effectively equivalent** in the sense that given any device of the four types considered, an equivalent device of any other type can be effectively constructed. Hence, we have established the following major result.

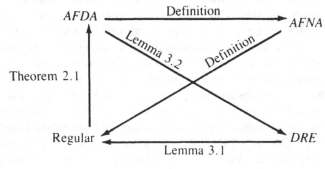

FIGURE 3.2

Theorem 3.3. *For an arbitrary language L, the following four conditions are effectively equivalent*: (i) *L is regular*, (ii) *L is AFDA*, (iii) *L is AFNA, and* (iv) *L is DRE*.

From now on, by Theorem 3.3 we always choose a representation for a regular language that is best suited to the purpose considered. In some cases, it will be more convenient to consider a finite deterministic automaton, whereas in some other cases, a regular expression will be more appropriate. A typical example is the proof of the next theorem, where important closure properties are established. Before stating the theorem, we introduce another operation that can be viewed as an extension of the operation of a morphism to subsets.

Definition 3.4. *For each letter a of an alphabet Σ, let $\sigma(a)$ be a language over an alphabet Σ_a. Further, define*

$$\sigma(\lambda) = \lambda, \qquad \sigma(uw) = \sigma(u)\sigma(w),$$

*for all words u and w over Σ, where the juxtaposition on the right side denotes catenation. Such a mapping σ of Σ^* into the set of languages over the union of the alphabets Σ_a is termed a **substitution**. For a language L over Σ, we define*

$$\sigma(L) = \{w \mid w \in \sigma(u) \text{ for some } u \in L\}.$$

*A family of languages is **closed under substitution** iff $\sigma(L)$ is in the family whenever L and each of the languages $\sigma(a)$ is in the family.*   □

Observe that morphisms are substitutions, where each of the languages $\sigma(a)$ consists of a single word.

Theorem 3.4. *The family of regular languages is closed under all the following operations: intersection, complementation, substitution (and, hence, under arbitrary morphisms), and mirror image.*

Proof. The complement of an AFDA language is obtained by interchanging the roles of final and nonfinal states in the corresponding FDA. This shows that the family of regular languages is closed under complementation. Since it is also closed under union, by Lemma 3.1, and since for all languages L_1 and L_2 we have

$$L_1 \cap L_2 = \sim(\sim L_1 \cup \sim L_2),$$

we conclude that the family of regular languages is closed under intersection.

Clearly, $\sigma(L)$ is always obtained by substituting every occurrence of a letter a by a regular expression for $\sigma(a)$ in the regular expression for L. (We, of course, assume that the languages involved are regular.) This shows closure under substitution.

Closure under mirror image is obtained in an analogous fashion: The mirror image of a language denoted by a regular expression is obtained by reversing the order of factors in all catenations in this regular expression. The resulting regular expression denotes the mirror image of the original language. □

The proof of Theorem 3.4 shows that the AFDA representation is suitable for Boolean operations (in fact, a direct proof can also be given for the closure under intersection), whereas the DRE representation is suitable for showing closure under substitution and mirror image. The reader is encouraged to consider how the proof could be carried out (if at all) with the roles of AFDA and DRE interchanged in regard to the operations in question.

We can now also strengthen our previous results concerning regular Post systems. It was shown in Lemma 2.9 that every PRPS language is regular. Clearly, an arbitrary RPS language L can be expressed in the form

$$L = L' \cap \Sigma_T^*,$$

where L' is PRPS and Σ_T is an alphabet. Since Σ_T^* is regular, Theorem 3.4 implies that L is regular. Hence, Theorem 2.10 can be strengthened to the following form.

Theorem 3.5.   *A language is regular iff it is PRPS iff it is RPS.*

Decision problems are of crucial importance in the theory of computing. Numerous decidability and undecidability results will be established in the sequel. We already have tools for establishing a number of decidability results for regular languages. Before establishing them, we introduce some terminology dealing with decision problems concerning languages.

Consider a class D of devices for defining languages. (Such a D might consist of all grammars of some type, all automata of some kind, or of some other devices, such as regular expressions.) By the **equivalence problem** for D we mean the problem of determining for any two devices d_1 and d_2 in D whether or not they define the same language. In the **inclusion problem** for D, we have to determine for arbitrary d_1 and d_2 in D whether or not the language defined by d_1 is included in the language defined by d_2. (Clearly, a solution for the inclusion problem yields a solution for the equivalence problem, whereas the inclusion problem may be undecidable and the equivalence problem still decidable.) By the **emptiness** (resp. **infinity**) **problem** for D we mean the problem of determining for an arbitrary device d in D whether or not the language defined by d is empty (resp. infinite). The **membership**

problem for D consists of deciding for an arbitrary d in D and for an arbitrary word w whether or not w is in the language defined by d.

Whenever we speak of problems such as those just given *for some specific language family* (for instance, about the emptiness problem for regular languages), then we consider some class D of devices defining exactly the language family in question. In case of regular languages, D might consist of one of the following: regular expressions, regular grammars, finite deterministic automata, finite nondeterministic automata, or regular Post systems. Since the devices are effectively equivalent (this also obviously holds true if regular Post systems are compared with the other devices), it is irrelevant which class of devices is considered from the point of view of decidability.

Theorem 3.6. The equivalence, inclusion, emptiness, infinity, and membership problems are all decidable for regular languages.

Proof. We consider the AFDA representation. (DRE representation is suitable for emptiness and infinity but not so suitable for the other problems; see Exercise 3.11.)

Thus, let $L = L(\text{FDA})$. To check the membership of a word w in L, we just have to run w on FDA. (If w contains letters not belonging to the input alphabet of FDA, then w is not in L.) To find out the nonemptiness (resp. infinity) of L, we check the existence of a path (resp. a path with a loop) from the initial state to one of the final states in the transition diagram for FDA. (An induction on the number of states provides a formal argument that this can be accomplished effectively.)

Further, let $L' = L(\text{FDA}')$. Clearly,

$$L \subseteq L' \quad \text{iff} \quad L - L' = \varnothing.$$

But $L - L'$ is regular by Theorem 3.4, and so its emptiness can be checked. (Observe that the proof of Theorem 3.4, as well as all of our proofs so far concerning closure under certain operations, are effective in the sense that an AFDA or some other representation is obtained. In this case, it is obtained for $L - L'$.) From the decidability of the inclusion problem, the decidability of the equivalence problem now readily follows. □

3.3. GENERALIZED SEQUENTIAL MACHINES

Finite automata with outputs will now be considered. Their main importance stems from the fact that the translation affected by them very often converts the input language into a form that is easier to handle. Moreover, important properties such as regularity are preserved in the translations.

In the original model of a sequential machine, the output was, at any given time instant, a letter determined by the input and state at that time instant. Our model of a generalized sequential machine defined below is more general. It is general enough for all our purposes. Information about a still more general finite automaton with outputs (called a rational transducer) is given in Exercise 3.13.

Definition 3.5. *A rewriting system* (Σ, P) *is called a* **generalized sequential machine** *iff the conditions* (*i*)–(*iii*) *are satisfied:*

(*i*) Σ *is divided into two disjoint alphabets,* Σ_Q *and* $\Sigma_I \cup \Sigma_O$. *The three sets* Σ_Q, Σ_I, *and* Σ_O *are called the* **state**, **input**, *and* **output** *alphabets, respectively. The sets* Σ_I *and* Σ_O *are nonempty but not necessarily disjoint.*

(*ii*) *An element* $q_0 \in \Sigma_Q$ *and a subset* $Q_F \subseteq \Sigma_Q$ *are specified and are referred to as the* **initial state** *and the* **final state set**.

(*iii*) *The productions in P are of the form*

$$q_i a \rightarrow w q_j, \qquad q_i, q_j \in \Sigma_Q; \quad a \in \Sigma_I; \quad w \in \Sigma_O^*. \tag{3}$$

Let GSM be a generalized sequential machine, $\Rightarrow$ *the direct-yield relation of the underlying rewriting system, and u a word over the input alphabet* Σ_I *of GSM. We let*

$$GSM(u) = \{w \mid q_0 u \Rightarrow^* w q_F \text{ for some } q_F \in Q_F\}.$$

Let L be a language over Σ_I. *Then GSM* **translates**, *or* **maps**, *L into the language*

$$GSM(L) = \{w \mid w \in GSM(u) \text{ for some } u \in L\}. \tag{4}$$

Mappings defined by (4) *are called* **gsm mappings**.

Similarly, if u is a word over the output alphabet Σ_O, *we define*

$$GSM^{-1}(u) = \{w \mid u \in GSM(w)\}.$$

For a language L over Σ_O, *we define*

$$GSM^{-1}(L) = \{w \mid w \in GSM^{-1}(u) \text{ for some } u \in L\}.$$

Mappings of this form are referred to as **inverse gsm mappings**.

GSM is **deterministic** *iff, for every* $q_i \in \Sigma_Q$ *and* $a \in \Sigma_I$, *there is exactly one production from* (3) *in P. A deterministic GSM is a* **Mealy machine** *iff every word w appearing in productions* (3) *consists of one letter of* Σ_O. □

Intuitively, productions (3) may be interpreted as follows. When scanning the letter a in the state q_i, one possible behavior for GSM is to expend a, output w, and go to the state q_j. Moreover, the whole translation can ultimately be rejected if a final state is not reached at the end. For a deterministic GSM, there is only one possible behavior in each configuration. The state transitions are similar as for a finite automaton (deterministic or non-deterministic)—only the output word w in (3) is additional. In fact, if our GSM is converted back to an acceptor by defining the accepted language to consist of such words u over Σ_I for which GSM(u) is nonempty, then the languages acceptable in this fashion are exactly the AFDA languages.

For a morphism h and a language L, the **inverse morphic image** of L is defined in the natural fashion:

$$h^{-1}(L) = \{w \mid h(w) \in L\}.$$

Observe that morphisms and inverse morphisms are special cases of gsm mappings and inverse gsm mappings. In fact, a deterministic GSM with only one state, q_0, can be associated with any morphism $h : \Sigma_I^* \Rightarrow \Sigma_O^*$. For every a in Σ_I, the production $q_0a \Rightarrow h(a)q_0$ is in GSM, and there are no other productions. (The state q_0 is the initial and the only final state.) It is clear that

$$h(L) = \mathrm{GSM}(L)$$

holds for all languages L over Σ_I.

Our main aim in this section is to show that gsm mappings, as well as inverse gsm mappings, preserve regularity. For future purposes, we present this result in a somewhat more general form.

We say that a family $\mathcal{L}$ of languages is a **cone** iff it is closed under morphisms, inverse morphisms, and intersection with regular languages. More explicitly, whenever L is in $\mathcal{L}$, h is a morphism and R is a regular language, then each of the languages $h(L)$, $h^{-1}(L)$, and $L \cap R$ is in $\mathcal{L}$. (We assume that the alphabets of h and L match in such a way that $h(L)$ is defined. To exclude trivialities, we also assume that $\mathcal{L}$ contains at least one nonempty language.)

It is easy to show that a cone always contains all regular languages R. (Indeed, since a cone $\mathcal{L}$ is closed under intersection with regular languages, a regular language $R \subseteq \Sigma_R^*$ is in $\mathcal{L}$ if Σ_R^* is in $\mathcal{L}$. Using the three closure properties, it is easy to see that Σ_R^* is in $\mathcal{L}$.)

We need two simple lemmas concerning regular languages. The first of them deals with languages L with the following property: A word w being or not being in L depends only on subwords of w of length 2. This is a special case of *locally testable* languages; see Exercise 3.16. The second lemma establishes another closure property.

Lemma 3.7. *Assume that Σ_1 is an alphabet and ρ is a binary relation defined on Σ_1. Then the language*

$$L_\rho = \{b_1 \cdots b_k \mid k \geq 1, \text{ each } b \text{ in } \Sigma_1, \text{ and } b_i \rho b_{i+1} \text{ holds}$$
$$\text{for all } i = 1, \ldots, k-1\}$$

is regular.

Proof.   Let $\Sigma_1 = \{a_1, \ldots, a_n\}$. Then the regular grammar with the production set

$$\{S \to a_i A_i \mid 1 \le i \le n\} \cup \{A_i \to a_j A_j \mid 1 \le i, j \le n \text{ and } a_i \rho a_j \text{ holds}\} \cup$$
$$\{A_i \to \lambda \mid 1 \le i \le n\}$$

generates the language L_ρ. $\square$

Lemma 3.8.   *The family of regular languages is closed under inverse morphisms.*

Proof.   Consider $L = L(\text{FDA})$ and an arbitrary morphism $h : \Sigma^* \to \Delta^*$. We assume without loss of generality that Δ is contained in the input alphabet Σ_T of FDA. (Otherwise, consider the restriction Δ' of Δ into Σ_T and the corresponding restriction of h, $h' : \Sigma'^* \to \Delta'^*$. Clearly, $h^{-1}(L) = h'^{-1}(L)$.) We construct a finite deterministic automaton FDA' as follows. The state alphabet and the initial state and final state sets of FDA' equal the corresponding items of FDA. The input alphabet of FDA' is Σ. The production $q_i a \to q_j$ is in FDA' iff

$$q_i h(a) \Rightarrow *_{\text{FDA}} q_j.$$

(Thus, for $h(a) = \lambda$, we have $i = j$.)

It is clear that the construct FDA' thus defined is a finite deterministic automaton. In particular, observe that the state q_j is uniquely determined for every pair (q_i, a).

It follows by the construction that an arbitrary word w is in $L(\text{FDA}')$ iff $h(w)$ is in $L(\text{FDA})$. But this means that

$$L(\text{FDA}') = h^{-1}(L).$$ $\square$

Theorem 3.9.   *Every cone is closed under gsm mappings.*

Proof.   Let $\mathcal{L}$ be a cone and L an arbitrary language in $\mathcal{L}$. Consider a GSM with the set P of productions of the form of (3). In regard to the different items of our GSM, we follow the notation introduced in the definition of GSM given above. We have to show that $\text{GSM}(L)$ is in $\mathcal{L}$.

Consider the auxiliary alphabet

$$\Sigma_1 = \{[q_i, a, w, q_j] \mid (3) \text{ is in } P\}.$$

Since P is finite, Σ_1 is clearly an alphabet. When referring to the letters b of Σ_1—that is, to quadruples $b = [q_i, a, w, q_j]$—we use expressions such as "the first state symbol in b," meaning q_i.

A binary relation ρ on Σ_1 is defined as follows: For any b and b' in Σ_1, $b \rho b'$ iff the second state symbol of b equals the first state symbol of b'. Then the language L_ρ as defined in Lemma 3.7 is regular.

Let L_1 be the subset of L_ρ consisting of all words u such that the first state symbol of the first letter of u is q_0, whereas the second state symbol of the last letter of u is in Q_F. Then, clearly,

$$L_1 = (K_1 \Sigma_1^* K_2 \cup K_3) \cap L_\rho,$$

where each K_i is the union of some letters of Σ_1 (maybe $K_i = \varnothing$). Consequently, L_1 is regular. Finally, we denote by L_2 the regular language $L_1 \cup \{\lambda\}$ or L_1, depending on whether or not q_0 is in Q_F.

Consider, finally, two morphisms h_1 and h_2 mapping Σ_1^* into Σ_I^* and Σ_O^*, respectively, defined by:

$$h_1([q_i, a, w, q_j]) = a, \qquad h_2([q_i, a, w, q_j]) = w.$$

It is now immediate that

$$\text{GSM}(L) = h_2(h_1^{-1}(L) \cap L_2). \tag{5}$$

By the closure properties of a cone, we conclude that $\text{GSM}(L)$ is in $\mathcal{L}$. □

Theorem 3.10. *Every cone is closed under inverse gsm mappings.*

Proof. Only a slight modification of the previous proof is needed. Let $\mathcal{L}$, L, and GSM be as in the previous proof. We have to show that $\text{GSM}^{-1}(L)$ is in $\mathcal{L}$.

We define Σ_1, ρ, L_ρ, L_1, L_2, h_1, and h_2 exactly as in the proof of Theorem 3.9. Then it is easy to see that

$$\text{GSM}^{-1}(L) = h_1(h_2^{-1}(L) \cap L_2). \tag{6}$$

Our theorem is now an immediate consequence of (6) and the definition of a cone. □

Equation (5), often referred to as **Nivat's theorem,** gives a normal form for gsm mappings. Every gsm mapping can be represented as the composition of three special gsm mappings—namely, inverse morphism, intersection with a regular language, and morphism.

The following theorem is now an immediate consequence of Theorem 3.4, Lemma 3.8, Theorem 3.9, and Theorem 3.10.

Theorem 3.11. *The family of regular languages is a cone. Hence, it is closed under gsm and inverse gsm mappings.*

In the definition of a cone, it is not required that the closure under the operations is effective—that is, that the result of an operation performed on a language L can effectively be constructed from a representation of L. However, our proofs show that whenever the closure under the cone operations actually is effective, then the closure established in Theorems 3.9 and 3.10 is also effective. Consequently, we may conclude that the family of regular languages is effectively closed both under gsm and inverse gsm mappings.

We conclude this section with an example of a general nature concerning applications of gsm mappings.

Example 3.3. Questions concerning number systems are very important both in cryptography and in coding theory. The customary **n-ary** notation, $n \geq 2$, of integers is based on words over the alphabet $\{0, 1, \ldots, n - 1\}$. A particular word $a_1 a_2 \cdots a_k$, where each a_i is a letter, denotes the integer

$$a_1 n^{k-1} + a_2 n^{k-2} + \cdots + a_{k-1} n + a_k. \tag{7}$$

Similarly, the **n-adic** notation, $n \geq 2$, is based on words over the alphabet $\{1, \ldots, n\}$, a particular word $a_1 a_2 \ldots a_k$ denoting the integer in (7). Thus, the *binary* word 10001 and the *dyadic* word 1121 both denote the same number, the number 17 in decimal (that is, 10-ary) notation.

Both *n*-ary and *n*-adic notations are **complete**: Every positive integer has a representation. However, the *n*-adic representation is *unambiguous* in the sense that every positive integer has a unique word representing it, whereas the *n*-ary representation is *ambiguous;* every positive integer has many (in fact, infinitely many due to arbitrarily long sequences of 0s) words representing it.   □

For many purposes it is useful to consider number systems generalized in the following two respects: (i) completeness is not required and (ii) the digits may be larger than the base. This leads us to the following definition. Since unambiguity is a desirable goal, the definition is based on generalizing the *n*-adic notation.

Definition 3.6. *A **number system** is a $(v + 1)$-tuple*

$$N = (n, m_1, \ldots, m_v)$$

*of positive integers such that $v \geq 1$, $n \geq 2$ and $1 \leq m_1 < m_2 < \cdots < m_v$. The number n is referred to as the **base** and the numbers m_i as **digits**. A nonempty word*

$$m_{i_0} m_{i_1} \cdots m_{i_{k-1}} m_{i_k}, \qquad 1 \leq i_j \leq v,$$

*over the alphabet $\{m_1, \ldots, m_v\}$ is said to **represent** the integer*

$$m_{i_0} + m_{i_1} n + \cdots + m_{i_{k-1}} n^{k-1} + m_{i_k} n^k.$$

The set of all represented integers is denoted by $S(N)$.   □

The following are typical important questions concerning number systems.

Which sets are of the form $S(N)$, for some number system N? When are two number systems equivalent in the sense that they represent the same set of numbers? When is a number system almost complete in the sense that all positive integers with a finite number of exceptions are represented by it? (The question of completeness—all integers are represented—is trivial.) When is a number system ambiguous; that is, when are there two distinct words representing the same number?

Such questions can be conveniently answered by considering gsm mappings. Observe that in our definition above we have reversed the customary notation in number systems: In our notation the leftmost digit makes the smallest (rather than the largest) contribution to the integer represented. This is irrelevant and only reflects the fact that our generalized sequential machines operate from left to right.

The basic tool for many problems concerning number systems is the following *translation lemma*. Consider a number system N with base n and digits $m_1, \ldots, m_v$. Then we can construct a GSM that translates every word of the form $m_{i_0} \ldots m_{i_k} \#$ (where the m's are digits and $\#$ is a new letter) into the word w over the alphabet $\{1, \ldots, n\}$ such that the number represented by $m_{i_0} \ldots m_{i_k}$ in N equals the number represented by w in reverse n-adic notation. Consequently, by Theorem 3.11, the set $S(N)$, viewed as a set of n-adic numbers, is regular.

Let $t = \max\{n, m_v\}$; then GSM is basically constructed as follows. The state set consists of the states $q_0, q_1, \ldots, q_{2t}$ and of a special final state q_F. Intuitively, being in the state q_i means that there is a carry i in the computation so far. Thus, when reading the input j in the state q_i, GSM produces the output letter j' and goes to the state $q_{i'}$, where i' and j' are unique integers satisfying

$$i + j = j' + i'n, \qquad 1 \le j' \le n.$$

Finally, when reading the letter $\#$ in the state q_i, GSM produces the output i in reverse n-adic notation and goes to the state q_F.

By the translation lemma, questions concerning number systems can be reduced to questions concerning n-adic notation. Many decision methods result in this fashion.

For instance, consider the number system $N = (2, 2, 3, 4)$. What is the set $S(N)$? We leave it for the reader to verify that $S(N)$ consists of all integers except the ones whose reverse dyadic representation is of the form 12^i for some $i \ge 0$. Thus, the missed numbers are of the form $2^k - 3$, for $k = 2$, $3, \ldots$, the first few ones being 1, 5, 13, 29, 61. This can be verified by answering the question: Which dyadic words can appear as outputs of GSM?

In the same way it can be verified that, for $N = (2, 1, 4)$, $S(N)$ equals the set of numbers incongruent to 2 modulo 3. Observe that N is unambiguous.

3.4. PUMPING LEMMAS

A natural way of showing that a given language is within a specific class of languages is to exhibit a proper device defining the language. For instance, to show that a given language L is context-free, we just have to construct a context-free grammar G such that $L = L(G)$.

It is in general much more difficult to show that a given language is *not* within a specific class of languages. In such a proof we have to consider all the (usually infinitely many) devices defining the languages in the class. However, there are many results to the effect that languages in a specific class have a certain property. Hence, a language not possessing that property cannot belong to the class.

In regard to regular and context-free languages, the most-useful of such negative criteria are known as *pumping lemmas*. This term reflects the following intuitive idea. Whenever a sufficiently long word w is chosen from the given language L, then a subword (or several subwords) of w can be exhibited such that if arbitrarily many copies of the subword are "pumped" next to it, the resulting word is always in L.

We first prove a standard version of the pumping lemma for regular languages (Theorem 3.12) and then the corresponding result for context-free languages (Theorem 3.13). We then strengthen the result concerning regular languages to yield a necessary and sufficient condition (Theorem 3.14).

Theorem 3.12. *For every regular language L, there is a natural number p such that every word x in L satisfying $|x| \geq p$ can be decomposed as $x = uvw$, where $v \neq \lambda$ and $uv^n w$ is in L for all $n \geq 0$. Moreover, u, v, and w can be chosen in such a way that $|uv| \leq p$. They can also be chosen in such a way that $|vw| \leq p$.*

Proof. We may assume that $L = L(\text{FDA})$ for some finite deterministic automaton FDA. We choose p to be the number of states of FDA. Consider any word $x = a_1 \cdots a_k$ in L, where the a's are letters and $k \geq p$. For $i = 1, \ldots, k$, let q_i be the state in which FDA is after reading the prefix $a_1 \cdots a_i$—that is, the unique state satisfying $q_0 a_1 \cdots a_i \Rightarrow^* q_i$, where q_0 is the initial state of FDA. Since x is in L, the state q_k is final. Moreover, by the choice of p, the states $q_0, q_1, \ldots, q_k$ are not all distinct. Let j be the smallest integer for which there is an integer $i < j$ such that $q_i = q_j$. (It is possible that $i = 0$ and/or $j = k$.) We choose $u = a_1 \cdots a_i$ (for $i = 0$, this reads $u = \lambda$), $v = a_{i+1} \cdots a_j$, and $w = a_{j+1} \cdots a_k$ (for $j = k$, this reads $w = \lambda$). Since $q_i = q_j$ and q_k is a final state, all words $x_n = uv^n w$, $n \geq 0$, are in L. Clearly, $|uv| \leq p$. Thus, the first two sentences of Theorem 3.12 have been established.

To establish the third sentence, we modify the construction above by letting i be the largest integer for which there is an integer $j > i$ such that $q_i = q_j$. For this pair (i, j), the words u, v, and w are defined exactly as before. Then the construction works as before, and now $|vw| \leq p$. This proves the third sentence.   □

Of course, in Theorem 3.12 we have to make a choice whether we want the beginning or the end of the words considered to be short—that is, whether the second or third sentence becomes applicable.

It is an immediate consequence of Theorem 3.12 that, for instance, the language $L = \{a^n b^n \mid n \geq 1\}$ is not regular. Indeed, what could the word v (having the pumping property) be? Such a word v cannot contain both of the letters a and b because then pumping would destroy the alphabetic order of letters in the words of L; all a's precede all b's. On the other hand, if v contains only one of the letters a and b, then pumping immediately destroys the equality between the number of occurrences of a and b in the words of L.

Theorem 3.13. *For every context-free language L, there are natural numbers p and q such that every word z in L satisfying $|z| > p$ may be written as $z = uvwxy$, where $vx \neq \lambda$ and $uv^n wx^n y$ is in L for each $n \geq 0$ and, moreover, $|vxw| \leq q$.*

Proof. If λ is in L, we determine the constants p and q for the language $L - \{\lambda\}$. Clearly, the same constants satisfy the required conditions for L. Hence, by Theorem 2.2, we may assume that $L = L(G)$, where G is in the Chomsky normal form. Let k be the number of nonterminals in G. We choose $p = 2^k$ and $q = 2^{k+1}$.

Let z be any word in L satisfying $|z| > p$. Considering an arbitrary derivation tree for z, we conclude that some path in the tree (from the root to one of the leaves) contains two occurrences of the same nonterminal A. This follows because some path must contain at least $k + 2$ nodes. For if the longest path contains a number of nodes less than or equal to $k + 1$, then $|z| \leq 2^k$. (In fact, this estimate is valid for all words z, terminal and nonterminal, derivable according to G. For terminal words, even the sharper estimate $|z| \leq 2^{k-1}$ holds.) On the other hand, only the last node in a path is labeled by a terminal letter.

Hence, z has a derivation according to G of the following form:

$$S \Rightarrow^* uAy \Rightarrow^* uvAxy \Rightarrow^* uvwxy = z.$$

From this, the pumping property immediately follows, since the derivation $A \Rightarrow^* vAx$ may be repeated an arbitrary number of times. Both of the terminal words v and x cannot be empty because G is in the Chomsky normal form. (One of them may be empty.) Hence, $vx \neq \lambda$.

Finally, to obtain the estimate $|vwx| \leq q$, we note that the two occurrences of A can be chosen in such a way that no path in the tree from the earlier A-labeled node to a leaf contains more than $k + 2$ nodes. (Intuitively, we consider a repetition as close to the leaves as possible. Since there are k

nonterminals and only leaves are labeled by terminals, a repetition must occur among the first $k + 2$ nodes, counting from the leaves). But now vwx is the word generated by the subtree whose root is the A-labeled node considered earlier. Hence, the estimate $|vwx| \leq 2^{k+1} = q$ follows.   □

The construction may be depicted as in Figure 3.3. Pumping means *pruning* the bigger A-subtree in place of the smaller A-subtree.

It is an immediate consequence of Theorem 3.13 that the language $\{a^n b^n c^n \mid n \geq 1\}$ is not context-free. Indeed, consider how pumping could be possible. Both of the words v and x must contain at most one of the letters a, b, or c. Otherwise, pumping destroys the alphabetic order of letters present in all words of the language. On the other hand, if both v and x contain at most one of the letters a, b, or c (and $vx \neq \lambda$), then pumping destroys the equality between the number of occurrences of a, b, or c.

It is easy to show (see Exercise 3.15) that the converses of Theorems 3.12 and 3.13 are not valid; the pumping property does not imply the language being regular or context-free. This is the reason why Theorems 3.12

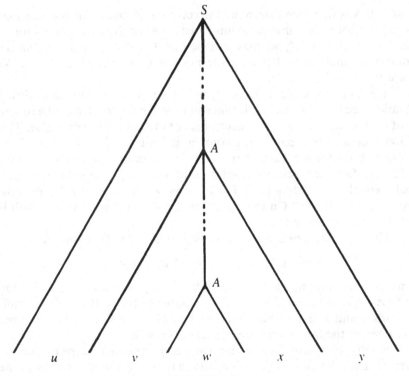

FIGURE 3.3

and 3.13 can be applied in the negative sense only—that is, to show that a given language is *not* regular or context-free. The next theorem gives a necessary and sufficient pumping condition for regular languages.

Theorem 3.14. *A language $L \subseteq \Sigma^*$ is regular iff there is a natural number p satisfying the following condition. Every word x over Σ of length p can be decomposed as $x = uvw$, where $v \neq \lambda$ and, for all words y over Σ and all $n \geq 0$, xy is in L exactly in the case $uv^n wy$ is in L.*

Proof. The *only if* part is established in the same way as Theorem 3.12. Let L be accepted by some FDA. We let p be the number of states of this FDA. Then every word x of length p causes a loop; that is, some state occurs twice. (See the proof of Theorem 3.12.) We now choose a decomposition $x = uvw$ in such a way that v is the word corresponding to this loop. Hence, for every word y and every $n \geq 0$, both of the words xy and $uv^n wy$ transfer the automaton from the initial state to the same state—that is,

$$q_0 xy \Rightarrow^* q_i \quad \text{and} \quad q_0 uv^n wy \Rightarrow^* q_j ,$$

where $q_i = q_j$. But this means that p satisfies the condition required.

Conversely, let $L \subseteq \Sigma^*$ be a language for which there is a natural number p satisfying the condition of our theorem. We construct a finite deterministic automaton FDA as follows. The input alphabet of FDA equals Σ. The state set consists of all elements q_z, where z ranges over all words over Σ with length less than or equal to $p - 1$. The initial state is q_λ. A state q_z specified by the word z (with $|z| < p$) is final iff z is in L.

Finally, the transition table of FDA is defined as follows: First consider a state specified by a word z with $|z| \leq p - 2$. Then every letter a of Σ causes the transition from q_z to q_{za}. Consider, secondly, a state specified by a word z of length $p - 1$ and a letter a of Σ. Since $|za| = p$, we may write $za = uvw$, where u, v, and w satisfy the condition of the theorem. Several such decompositions may be possible for za. Since we are constructing a finite deterministic automaton, we want the decomposition considered to be unique. For this purpose, we assume that uv is the shortest possible word and, moreover, that v is the shortest possible word within this uv. More specifically, there are no words u', v', and w' satisfying the condition of Theorem 3.14 such that $za = u'v'w'$ and $|u'v'| < |uv|$, and no such words u', v', and w' with the properties $|u'v'| = |uv|$ and $|v'| < |v|$. We now agree that the letter a causes the transition from the state q_z to the state q_{uw}. This completes the definition of FDA. (Observe, intuitively, that for short words we have a complete tree of transitions: Two distinct words lead the automaton to distinct states from the initial state. Only from the leaves—that is, from states specified by words of length $p - 1$—do we loop back to previous states.)

Having completed the definition of FDA (clearly, the construct is a finite deterministic automaton!), we still have to show that $L = L(\text{FDA})$. Assume the contrary, and let x be a word belonging to exactly one of the languages L and $L(\text{FDA})$. Moreover, we assume that x is the shortest among such words; that is, there is no word x' belonging to exactly one of the languages L and $L(\text{FDA})$ such that $|x'| < |x|$.

It is immediate from the construction of FDA that $|x| \geq p$. Hence, we may write $x = x_1 y$, where $|x_1| = p$. We now make use of the condition characterizing p and write $x_1 = uvw$, where u, v, and w satisfy the condition. We assume that $|uv|$ and $|v|$ are minimal in the same sense as in the definition of FDA. Hence, $x = uvwy$ is in L iff $x' = uwy$ is in L.

On the other hand, both the words uvw and uw cause the state transition from the initial state to the state q_{uw}. For the word uw this follows by the transition rules for short words and for the word uvw, by the transition rules for words of length $p - 1$. Consequently, both the words x and x' cause a state transition from the initial state to the same state. But this means that x is in $L(\text{FDA})$ iff x' is in $L(\text{FDA})$.

We now conclude that x' belongs to exactly one of the languages L and $L(\text{FDA})$. Since clearly $|x'| < |x|$, this contradicts the choice of x. Therefore, $L = L(\text{FDA})$, implying that L is regular. $\square$

3.5. PUSHDOWN AUTOMATA

The strictly finitary model of computation studied so far in this chapter remains unsatisfactory for most purposes. It is not capable of accepting even such a simple language as $\{a^n b^n \mid n \geq 1\}$ because it is not able to count the number of a's scanned before the first b is encountered. On the other hand, the straightforward generalization of FDA obtained by allowing infinitely many states is clearly not a satisfactory model of computation. It is too general and does not satisfy the essential requirement of effectiveness. Indeed, it is easy to see (Exercise 3.8) that every language whatsoever is acceptable by such an infinite automaton.

The general model of computation will be investigated in the next chapter. Numerous other models whose capabilities lie between those of the general model (Turing machine) and those of a finite automaton have been introduced. In most cases, such models are obtained by adding a potentially infinite working tape to a finite automaton. The resulting automata can be classified by imposing restrictions on the fashion of retrieving information from the working tape.

A natural method of retrieving information from the working tape is based on the principle "first in–last out" (or "last in–first out"). Such a working tape is referred to as a **pushdown tape**. Thus, the mode of operation is the same as that of a pushdown stack used in programming. Indeed, from the point of view of applications—in particular, constructions dealing with parsers—the resulting *pushdown automata* are most important.

Pushdown automata are discussed in this section. Since the detailed theory of context-free languages and the theory of parsing lie outside the scope of this book, our discussion about pushdown automata is restricted to only the most basic facts.

We begin with an intuitive description of a pushdown automaton. A pushdown automaton is a finite automaton combined with a potentially infinite pushdown tape. More specifically, this means the following.

A pushdown automaton reads words w over a fixed-input alphabet Σ_T letter by letter in discrete steps. Its actions also depend on the contents of the pushdown tape. Initially, the pushdown tape contains the letter z_0, and the automaton reads the first letter of w in the *initial* state.

At each moment of time during the computation, the automaton reads some letter a of w in some internal state q and has some letter z topmost (rightmost) on the pushdown tape. Its action, depending on the triple (a, q, z), consists of a state transition to a state q' and of replacing the letter z by some word, including the empty word, over the *pushdown alphabet*—that is, the alphabet used on the pushdown tape. The replacement of z by the empty word corresponds to "popping up" the contents of the pushdown tape below z.

The basic model of a pushdown automaton is *nondeterministic:* there may be several (but always finitely many) possible actions caused by a triple (a, q, z). Moreover, in such a triple a may equal the empty word (instead of being a letter of Σ_T). Such λ-**moves** make it possible for the automaton to change its state and also the contents of the pushdown tape without advancing the input tape at all.

In the basic model, a word w is accepted iff w gives rise to a computation leading to a final state. (As in connection with finite automata, a subset of states called *final states* has been specified in the definition.) The acceptance is independent of the final contents of the pushdown tape. Since the automaton is nondeterministic, the word w might give rise to several computations. The word w is accepted iff at least one of these computations is successful, independently of the possible failures caused by w.

Let us repeat what a successful computation means. Initially, the automaton scans the first letter of w in the specified initial state and has the specified *start letter* z_0 on the pushdown tape. At each moment of time, the scanned symbol a (either a letter of Σ_T or the empty word), the current state q, and the topmost (rightmost) letter z on the pushdown tape produce, nondeterministically, a pair (q', u). (In other words, several such pairs may be produced.) This means that the automaton replaces z by u on the pushdown tape, changes its state to q', and consumes a from the input tape. The word w is accepted iff there is a computation, after which the entire word w has been consumed and the automaton is in a final state.

We are now ready for the formal definitions.

Definition 3.7.   *A rewriting system (Σ, P) is called a **pushdown automaton** (PDA) iff conditions (i)–(iii) are satisfied:*

  (i) *Σ is divided into two disjoint alphabets Σ_Q and $\Sigma_T \cup \Sigma_{PD}$. The sets Σ_Q, Σ_T, and Σ_{PD} are called the **state**, **input** (or **terminal**), and **pushdown alphabet**, respectively. The sets Σ_T and Σ_{PD} are nonempty but not necessarily disjoint.*

 (ii) *Elements $q_0 \in \Sigma_Q$ and $z_0 \in \Sigma_{PD}$ and a subset $Q_F \subseteq \Sigma_Q$ are specified, referred to as the **initial state**, **start symbol**, and **final state set**, respectively.*

(iii) *The productions in P are of the two forms*

$$zq_i \to uq_j, \qquad z \in \Sigma_{PD}; \quad u \in \Sigma_{PD}^*; \quad q_i, q_j \in \Sigma_Q, \qquad (8)$$

$$zq_i a \to uq_j, \qquad z \in \Sigma_{PD}; \quad u \in \Sigma_{PD}^*; \quad a \in \Sigma_T; \quad q_i, q_j \in \Sigma_Q. \qquad (9)$$

*The language **accepted** by a pushdown automaton PDA is defined by*

$$L(PDA) = \{w \in \Sigma_T^* \mid z_0 q_0 w \Rightarrow^* u q_F \text{ for some } u \in \Sigma_{PD}^* \text{ and } q_F \in Q_F\}. \quad (10)$$

*A language L is **acceptable** by a pushdown automaton, briefly **APDA**, iff there is a PDA such that $L = L(PDA)$.*

*A pushdown automaton is **deterministic** iff for each $q_i \in \Sigma_Q$ and $z \in \Sigma_{PD}$, P contains either exactly one production from (8) and no productions from (9) or else no productions from (8) and exactly one production from (9) for every $a \in \Sigma_T$. Languages acceptable by deterministic pushdown automata, DPDA's, are referred to as **ADPDA** languages.*   □

Productions of the form of (8) may be interpreted to mean that after being in state q_i with z topmost on the pushdown tape, PDA may go to p_j and replace z by u without advancing the input tape at all. Productions (9) mean that after scanning the input letter a in state q_i with z topmost on the pushdown tape, PDA may go to q_j, replace z by u, and "consume" a from the input tape. (After this a can never be recovered; that is, the input tape moves in one direction only.)

For DPDA's, there is only one possible behavior at each moment of time. More specifically, the pair (z, q_i) determines first whether λ or some letter a is read from the input tape. If λ is read, the new contents of the pushdown tape, as well as the new state, are uniquely determined. If a is read, the triple (z, q_i, a) determines uniquely the new contents of the pushdown tape and the new state. ADPDA languages are also called **deterministic** languages for simplicity.

In what follows, we list some basic results concerning PDA's. The proofs of Theorems 3.15–3.17 are omitted. The proofs are, essentially, simulation arguments concerning how some device or mode of acceptance can be simulated by another. The ideas underlying the proofs are, thus, rather simple. On the other hand, quite a bit of technical apparatus is needed for detailed proofs. The matrix approach described briefly below simplifies these and similar proofs considerably.

The first result shows that pushdown automata bear the same relation to context-free languages as finite automata do to regular languages. However, deterministic PDA's define a strictly smaller family of languages.

Theorem 3.15. *A language is APDA iff it is context-free. Deterministic languages constitute a proper subfamily of context-free languages.*

The next theorem can be viewed as a "normal-form result" for pushdown automata; productions (8)—that is, λ-moves—are not necessary.

Theorem 3.16. *Every language acceptable by a pushdown automaton is acceptable by a pushdown automaton having no productions of the form of (8).*

The notion of acceptance used in defining equation (10) is customarily referred to as acceptance by **final state**. This means that a word is accepted iff it gives rise to a computation leading to a final state; the final contents of the pushdown tape are ignored. Acceptance by the **empty pushdown tape** is defined as the term indicates: A word is accepted iff it gives rise to a computation ending with an empty pushdown tape. Thus, in this mode of acceptance, the state of the automaton at the end of the computation is irrelevant. (Hence, the specification of the set Q_F also becomes unnecessary.) Acceptance of a word by **final state and the empty pushdown tape** combines the two modes mentioned: a word is accepted iff it gives rise to a computation, after which the automaton is in a final state *and* the pushdown tape is empty.

Thus, in addition to the language $L(PDA)$ as defined in (10), we may consider the following two languages accepted by PDA, corresponding to the other two modes of acceptance:

$$N(PDA) = \{w \in \Sigma_T^* \mid z_0 q_0 w \Rightarrow * \; q_i \text{ for some } q_i \in \Sigma_Q\}, \tag{11}$$

$$LN(PDA) = \{w \in \Sigma_T^* \mid z_0 q_0 w \Rightarrow * \; q_F \text{ for some } q_F \in Q_F\}. \tag{12}$$

However, the three different modes of acceptance yield the same family of acceptable languages.

Theorem 3.17. *A language L is APDA iff there is some PDA such that $L = N(PDA)$ iff there is some PDA such that $L = LN(PDA)$.*

By Theorems 3.15 and 3.17, a language is context-free iff it is acceptable by a pushdown automaton, no matter which of the three modes of acceptance is considered.

On the other hand, the three modes of acceptance cause also definite differences. For instance, if deterministic pushdown automata are considered, then (10) and (11) do not give rise to the same family of languages. Indeed, in this case, languages of the form in (11) constitute a proper subfamily of languages of the form in (10). Every language L of the form in (11), where PDA is deterministic, has the property that no proper prefix of a word in L is in L. This property is not necessarily possessed by languages of the form in (10), although PDA is deterministic.

It is easy to see that Theorem 3.16 is not valid in the deterministic case: λ-moves are necessary for (some) deterministic pushdown automata. This will be discussed further in the following example.

Example 3.4.   We begin with a pushdown automaton for the language $\{a^n b^n \mid n \geq 1\}$. This is a very typical language for PDA-acceptance: The a's are first pushed down. After the appearance of the first b on the input tape, the a's are popped up against the b's. If the b's exactly exhaust the a's, the word is accepted.

The formal definition of such a PDA is now given. The state, input, and pushdown alphabets are

$$\{q_0, q_1, q_2\}, \qquad \{a, b\}, \qquad \{z_0, a\},$$

respectively. The initial state is q_0, and q_2 is the only final state. The start symbol is z_0. Finally, the productions are:

$$z_0 q_0 a \rightarrow z_0 a q_0, \qquad a q_0 a \rightarrow a a q_0,$$
$$a q_0 b \rightarrow q_1, \qquad a q_1 b \rightarrow q_1,$$
$$z_0 q_1 \rightarrow q_2.$$

It can immediately be verified that the pushdown automaton thus defined satisfies the equations

$$L(\text{PDA}) = N(\text{PDA}) = LN(\text{PDA}) = \{a^n b^n \mid n \geq 1\}.$$

Our PDA is not deterministic according to the definition given above, the reason being that there are no productions for some configurations—for instance, there is no production whose left side is $z_0 q_0 b$. However, such productions can easily be added without altering the set of accepted words. Hence, our PDA is essentially deterministic.

The reader might want to construct a DPDA accepting the language

$$L_1 = \{wc\,\text{mi}(w) \mid w \in \{a, b\}*\}.$$

(Hence mi denotes mirror image, an operation defined in Example 2.6.) The basic idea in the construction is clear: w is pushed down until c is met, after which letters on the pushdown tape are popped up against letters on the input tape.

Also, the language

$$L_2 = \{w\,\mathrm{mi}(w) \mid w \in \{a, b\}*\}$$

is acceptable by a pushdown automaton. The automaton starts in the "pushing mood." At any moment of time, it has the option of either continuing in the pushing mood or switching into the "popping mood." It has to guess when the right moment has come—that is, when the middle of the input has been reached. There is no such problem in connection with L_1 because c indicates the middle of the input. It is intuitively clear that there is no way to eliminate guessing from the acceptance of L_2. The intuitive idea can be formalized: L_2 is not deterministic. Languages such as L_2 show that the second sentence of Theorem 3.15 is true.

We still mention the following two languages to illustrate further the importance of λ-moves and the mode of acceptance in connection with **deterministic** pushdown automata. The language

$$\{a^m b^n c^p a^m \mid m \geq 1, n \geq p \geq 1\}$$

is deterministic but not acceptable by a DPDA that does not use λ-moves. The language

$$\{a^m b^n \mid m \geq n \geq 1\}$$

is acceptable by a DPDA not using λ-moves but not acceptable by such a device by the empty pushdown store. ☐

We shall not enter a detailed discussion on *decision problems* concerning pushdown automata. Because in Theorem 3.15 the transition from a pushdown automaton to a context-free grammar (and vice versa) is effective, we conclude that PDA decision problems may be identified with decision problems for context-free grammars—at least as far as the defined languages are concerned. It will be seen in Chapter 5 that the equivalence problem for context-free languages is undecidable. Whether or not the same holds true for deterministic languages is one of the most celebrated open problems in automata theory.

We now briefly describe a technique of using matrices in the definition of and constructions dealing with pushdown automata. The technique is particularly useful in proofs because it makes vague simulation-oriented arguments computational: The whole proof becomes a computation involving mainly matrix products. The reader is referred to [Ku] for more details.

The basic idea is that all information about a pushdown automaton can be expressed in a single matrix. The matrix describes what can happen in one step of a computation, the product of two matrices what can happen in two steps, and so forth.

Let us discuss first how this can be accomplished for finite automata. Consider the automaton of Example 3.2. The state transitions can be expressed by the following 4×4 matrix:

$$M = \begin{pmatrix} \varnothing & \varnothing & a & b \\ b & \varnothing & \varnothing & a \\ a & \varnothing & \varnothing & b \\ b & a & \varnothing & \varnothing \end{pmatrix}.$$

More specifically, it is assumed that the rows and columns of the matrix are indexed with the states q_0, q_1, q_2, and q_3 (in this order). The entries of the matrix indicate the letters causing the corresponding state transition. For instance, the letter b appears in the entry (q_2, q_3) because b causes a state transition from q_2 to q_3. The matrix can be viewed as the adjacency matrix of the graph of the automaton. In general, the entries consist of the empty set, $\varnothing$, and of finite sets of letters.

The entry (q_0, q_0) in M^2 consists of a^2 and b^2. This tells us that a^2 and b^2 are the words causing a state transition from q_0 to q_0 in two steps. Similarly, because aba is the only word in the entry (q_0, q_1) of M^3, we conclude that aba is the only word causing a state transition from q_0 to q_1 in three steps.

The matrix M^*, essentially the union of all powers of M, can be defined and computed in the natural fashion. The language accepted by the automaton is the union of the proper entries in the row indexed by the initial state q_0 in the matrix M^*. Of course, the "proper" entries are obtained from columns indexed by final states—in our example, from the first two columns.

We are now ready to describe pushdown automata in a similar fashion. The rows and columns of the matrix M are now indexed with pairs (q, u), where q is a state and u is a word over the pushdown alphabet. Thus, the matrix M will be infinite. However, only finitely many entries in each row and column of M will be nonempty. This guarantees that the powers M^i exist.

The entry $((q_1, u_1), (q_2, u_2))$ indicates the letters x such that it is possible for the automaton to go from the state q_1 and pushdown-tape contents u_1 to the state q_2 and pushdown-tape contents u_2 when reading x from the input tape. Here x can be also the empty word. Thus, the entries of M consist of the empty set $\varnothing$ and of finite languages whose words are of length less than or equal to 1.

Furthermore, it is assumed that M satisfies the following **pushdown condition.** All rows and columns of M have only finitely many nonempty entries. Moreover, an entry $((q_1, u_1), (q_2, u_2))$ always equals the entry $((q_1, z), (q_2, u_3))$ if there are words

$$z \in \Sigma_{PD}, \quad u_4 \in \Sigma_{PD}^* \qquad \text{such that} \qquad u_1 = u_4 z \quad \text{and} \quad u_2 = u_4 u_3.$$

If no such words z and u_4 exist, then the entry $((q_1, u_1), (q_2, u_2))$ equals $\varnothing$.

The pushdown condition guarantees that only the topmost letter on the pushdown tape is used to determine the behavior.

The alphabets, initial and final states, start symbol, and a matrix M satisfying a pushdown condition determine a pushdown automaton completely. A pushdown matrix is always finitely specified. In fact, it is completely determined by the entries corresponding to the productions of the automaton. The reader might want to verify that the matrix product works as intended; the entries of M^2 indicate what happens on the input tape during two computation steps. The notion of acceptance most conveniently associated to the matrix representation is that of acceptance by final state *and* the empty pushdown tape—that is, the notion corresponding to (12).

For instance, in regard to the automaton PDA discussed at the beginning of Example 3.4, some of the entries are

$$((q_0, z_0), (q_0, z_0 a)) : a, \qquad ((q_0, a), (q_0, aa)) : a,$$
$$((q_0, a), (q_1, \lambda)) : b, \qquad ((q_1, a), (q_1, \lambda)) : b$$
$$((q_1, z_0), (q_2, \lambda)) : \lambda.$$

By the pushdown condition we infer that, for instance, the entry $((q_0, aaz_0), (q_0, aaz_0a))$ consists of a. Moreover, the entries listed above use the pushdown condition to determine all the nonempty entries of the matrix.

We mention, finally, a special case and a generalization of pushdown automata. The special case, referred to as a **counter automaton,** is obtained by assuming that the pushdown alphabet consists of one letter only. The resulting **counter languages** form a widely studied subfamily of context-free languages. For counter automata, the families of languages of the form in (10) and (11) are incomparable.

A natural generalization of a pushdown automaton is obtained by allowing more than one pushdown tape. It turns out that two pushdown tapes are enough in the sense that a device with more than two tapes can be simulated by one with two tapes. In fact, an automaton with two pushdown tapes possesses the capacity of the most general computing device, namely, the Turing machine discussed in Chapter 4.

EXERCISES

3.1. Consider the operations introduced in Section 2.1. Under which operations is the family of regular (resp. context-free) languages closed? In particular, show that the family of context-free languages is not closed under intersection; that is, the intersection of two context-free languages is not necessarily context-free. Prove that, on the other hand, the intersection of a context-free and a regular language is always context-free. The reference [Sa3] contains a survey of closure properties of language families.

3.2. Prove that the family of recursively enumerable languages is closed under the operations considered in Lemma 3.1. In regard to catenation and catenation closure, be careful that the different derivations are not mixed up in your construction. (This exercise can be viewed as a special case of the previous one.)

3.3. Show that if L is a regular language, then so is the following language L_1, consisting of the first halves of the words in L:

$$L_1 = \{w \mid \text{for some } w', \; ww' \in L \text{ and } |w| = |w'|\}.$$

Also, prove that if L is regular, then so is the language consisting of the middle third of each word in L.

3.4. Define the **weight** of a regular language, in symbols, weight(L), to be the smallest number of states in a finite deterministic automaton accepting L. Construct an algorithm for finding weight(L), as well as the corresponding minimal automaton. Show that the latter is unique up to isomorphism. (The rather trivial algorithm based on an exhaustive search is in most cases impossible in practice. See [Sa2] for a more practical algorithm.)

3.5. Construct, for every sufficiently large positive integer k, a regular language L_k such that L_k is accepted by a finite nondeterministic automaton with less than weight(L_k)$/k$ states.

3.6. Define the **representation number** of a regular language L, in symbols repr(L), to be the smallest possible number of final states in a finite deterministic automaton accepting L. Show by examples that repr(L) can be arbitrarily large.

3.7. Let L be a language over the alphabet Σ. The **equivalence relation** $\equiv_L$ induced by L is the binary relation on the set Σ^* defined by the following condition: $u \equiv_L v$ holds iff there is no word w such that exactly one of the words uw and vw is in L. Show that L is regular iff the index (the number of equivalence classes) of $\equiv_L$ is finite. Show also that weight(L) equals the index of $\equiv_L$. How can repr(L) be determined from $\equiv_L$?

3.8. Show that every language whatsoever is accepted by an infinite automaton (the device obtained from FDA by allowing infinitely many states). Generalize the notion repr(L) to concern arbitrary languages. Give examples of languages L such that repr(L) is infinite. Characterize the set of languages L such that repr(L) is finite.

3.9. Prove that L^* is regular if L is any language whatsoever over the one-letter alphabet $\{a\}$.

3.10. Regular grammars are often called **right-linear** because of the form of the productions. Define the notion of a **left-linear** grammar and show that left-linear grammars generate the same family of languages as right-linear ones. (*Hint:* The mirror image of a regular language is regular.)

3.11. Prove directly (by considering only regular expressions and not at all automata) that emptiness and infinity problems are decidable for DRE languages. Why is this direct approach not suitable for proving the other parts of Theorem 3.6?

3.12. Show that membership, emptiness, and infinity problems are decidable for context-free languages. (In regard to the infinity problem, Theorem 3.13 is of crucial importance.)

3.13. A **rational transducer** (sometimes also called an *a-transducer*) is defined as a generalized sequential machine except that in productions (3), a can be an arbitrary word over Σ_I. (Thus, a transducer may read an arbitrary word, including the empty word, from the input tape during one computation step.) The corresponding mappings are called **rational transductions**. Prove that every cone is closed under rational transductions.

3.14. Prove that a family of languages is closed under rational transductions iff it is closed under morphisms, inverse morphisms and intersections with regular languages. (Show that the three operations mentioned are rational transductions and that every rational transduction can be expressed in terms of the three operations.) Which operations correspond in the same sense to gsm mappings?

3.15. Show by examples that the converses of Theorems 3.12 and 3.13 are not valid—that is, that the pumping property does not guarantee that the language is regular or context-free.

3.16. For $k \geq 2$, a regular language L is termed **k-testable** iff the set of subwords of length k obtained from a word $\#w\#$ uniquely determines whether or not w is in L. (In other words, if w_1 and w_2 yield the same set of subwords of length k, then w_1 is in L iff w_2 is in L. The marker $\#$ is used to indicate the beginning and end of w.) A language is **locally testable** iff it is k-testable for some k. Show that the language L in Example 2.2 is locally testable. Give examples of regular languages that are not locally testable. (A general result here is that every locally testable language can be expressed without the star—that is, in terms of only Boolean operations and catenation, starting from the atomic languages.)

3.17. Prove that if $v < n$ in a number system N, then no number system N_1 whose base differs from n is equivalent to N. (The notation here is the same as in Example 3.3.)

CHAPTER 4

Turing Machines and Recursive Functions

4.1. A GENERAL MODEL OF COMPUTATION

As is true for all our models of computation, a *Turing machine* also operates in discrete time. At each moment of time it is in a specific internal (memory) state, the number of all possible states being finite. A read-write head scans letters written on a tape one at a time. A pair (q, a) determines a triple (q', a', m) where the q's are states, a's are letters, and m ("move") assumes one of the three values l (left), r (right), or 0 (no move). This means that, after scanning the letter a in the state q, the machine goes to the state q', writes a' in place of a (possibly $a' = a$, meaning that the tape is left unaltered), and moves the read-write head according to m.

If the read-write head is about to "fall off" the tape, that is, a left (resp. right) move is instructed when the machine is scanning the leftmost (resp. rightmost) square of the tape, then a new blank square is automatically added to the tape. This capability of indefinitely extending the external memory can be viewed as a built-in hardware feature of every Turing machine. The situation is depicted in Figure 4.1.

The input-output format of a Turing machine is specified as follows. The machine begins its computation by scanning the leftmost letter of a given input word in a specific initial state. The input is accepted iff the computation reaches a specific final state. If the machine is viewed as a translator

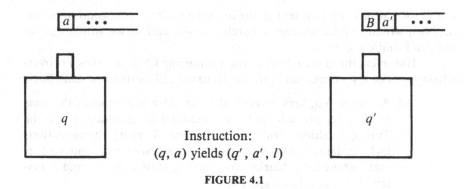

Instruction:

(q, a) yields (q', a', l)

FIGURE 4.1

rather than an acceptor, then the word on the tape, after the machine has reached a final state, constitutes the output to the given input. Some of the letters on the tape might thus be disregarded.

Thus, the tape can be viewed both as an input and output channel and as a potentially infinite memory. Basic differences between Turing machines and other types of automata can be briefly described as follows. A finite automaton has only an internal memory determined by its finite state set; the input tape is not used as an additional memory. A finite automaton just reads the input in one sweep from the left to the right. In a linear-bounded automaton, the external memory is bounded from above by the size of the input word (or by a linear function of it, which amounts to the same thing). We do not discuss linear-bounded automata in this book. In a pushdown automaton, the access to the information in the external memory is very limited.

Hence, clearly, a Turing machine is more general than the other models of computation we have considered. It is also *the general* model: Every algorithm (in the intuitive sense) can be realized as a Turing machine. Thus, if you have an algorithm for computing the values $f(x)$ of a function f given the argument value x, then you may construct a Turing machine producing the output $f(x)$ from the input x for all x. Or, if you have an effective procedure for listing all words of a language L, then you may construct a Turing machine that accepts a word iff it is in L. The statement that a Turing machine is a general-enough mathematical model for the intuitive notion of an effective procedure is customarily referred to as **Church's thesis.** Clearly, each Turing machine constitutes an effective procedure in the intuitive sense—so this mathematical formalization is not too general. Church's thesis asserts that the formalization is not too narrow, implying that it is exactly the right one.

In order to prove Church's thesis, we would have to compare effective procedures (an intuitive notion) and Turing machines (a formal mathematical notion). Consequently, we would have to formalize the notion of an effective procedure. But we would then again face this problem: Is the introduced formalization equivalent to the intuitive notion? The solution of this problem would require another Church's thesis, and so we would end up with an infinite regression.

However, the empirical evidence supporting Church's thesis is overwhelming. The arguments can be divided into the following three main lines:

(i) All suggested, very diverse, alternate formalizations of the class of effective procedures have turned out to be equivalent to the Turing machine formalization. These alternate formalizations include the rewriting systems (with a suitable input-output format), grammars, Markov algorithms, Post systems, and L systems discussed in Chapter 2.

(ii) The class of mappings computable by Turing machines has very strong invariance properties. For instance, all extensions considered have been shown equivalent to the original Turing machine model. The extensions include machines with more than one tape and/or read-write head (there may even be unboundedly many of them!), as well as machines working nondeterministically or in dimensions higher than one. The invariance properties also correspond to the intuitive idea that if we have effective procedures for solving certain problems, then we also have effective procedures for problems resulting from the given ones by "combining" them in a reasonably effective way. Thus, if two functions $f_1(x)$ and $f_2(x)$ are effectively computable, then so are their product $f_1(x) \cdot f_2(x)$ and their composition $f_1(f_2(x))$.

(iii) All intuitively effective procedures (for computing functions, accepting languages, and so on) *considered so far* have found their formal counterpart; that is, their "programming" on Turing machines has turned out to be possible. This fact was referred to by Markov as the **normalization principle** of algorithms. Of course, the normalization principle extended to all intuitively effective procedures is equivalent to Church's thesis.

We have been speaking of Turing machines as a formal mathematical notion without actually giving the details of the formalization. The reader might already have been able to find out the details on the basis of our informal remarks above. The following definition gives a formalization in terms of rewriting systems. If only Turing machines are considered, this formalization is not the most elegant one. However, it is the shortest in terms of the formal apparatus developed so far.

Definition 4.1. *A rewriting system* $TM = (\Sigma, P)$ *is called a* **Turing machine** *iff conditions* (i)–(iii) *are satisfied.*

(i) Σ *is divided into two disjoint alphabets* Σ_Q *and* Σ_O, *referred to as the* **state** *and the* **operating** *alphabets, respectively.*

(ii) *Elements q_0 and q_F of Σ_Q are specified and are referred to as the* **initial** *and the* **final** *states. (The equality $q_0 = q_F$ is allowed.) Distinct elements # and B of Σ_O (referred to as the* **boundary marker** *and the* **blank symbol**) *are specified. Furthermore, it is assumed that the set $\Sigma_1 = \Sigma_O - \{\#, B\}$ is not empty, and a subset $\Sigma_T \subseteq \Sigma_1$ (the* **terminal** *alphabet) is specified.*

(iii) *The productions in P are of the forms*

$$q_i a \rightarrow q_j b \qquad (overprint), \tag{1}$$

$$q_i ac \rightarrow aq_j c \qquad (move\ right), \tag{2}$$

$$q_i a\# \rightarrow aq_j B\# \qquad (move\ right\ and\ extend\ workspace), \tag{3}$$

$$cq_i a \rightarrow q_j ca \qquad (move\ left), \tag{4}$$

$$\#q_i a \rightarrow \#q_j Ba \qquad (move\ left\ and\ extend\ workspace), \tag{5}$$

where $q_i \in \Sigma_Q - \{q_F\}$, $q_j \in \Sigma_Q$, and a, b, and c are in $\Sigma_1 \cup \{B\}$. Furthermore, for each q_i in $\Sigma_Q - \{q_F\}$ and a in $\Sigma_1 \cup \{B\}$, P either contains no productions (2) and (3) (resp. (4) and (5)) or else contains both (3) and (2) for every c in $\Sigma_1 \cup \{B\}$ (resp. contains both (5) and (4) for every c in $\Sigma_1 \cup \{B\}$). For no q_i in Σ_Q and a in $\Sigma_1 \cup \{B\}$, the word $q_i a$ is a subword of the left side in two productions of the forms (1), (3), and (5).

The language accepted by Turing machine TM is defined by

$$L(TM) = \{w \in \Sigma_T^* \mid \#q_0 w\# \Rightarrow^* \#w_1 q_F w_2\# \text{ for some words } w_1 \text{ and } w_2\}.$$

A Turing machine **halts** *with a word w if there are words w_1 and w_2 such that*

$$\#q_0 w\# \Rightarrow^* \#w_1 q_F w_2\#.$$

Otherwise, it **loops** *with w.*   □

Thus, halting means reaching the final state, q_F, for which there are no instructions according to our definition. Of course, the computation may also reach a configuration (q_i, a) such that $q_i \neq q_F$ and there is no production where $q_i a$ occurs as a subword on the left side. In this case, according to our definition, *TM* loops with the input. Indeed, in this case we may add productions causing a loop without affecting the output associated with a word or the acceptance of a word.

The reader is encouraged to compare our formal definition with the previously given intuitive explanations. If a word $\#w' q_i w''\#$ has been derived according to the rewriting system from the word $\#q_0 w\#$, this means that the computation from the input w has reached the point where *TM* scans the first letter of w'' in the state q_i, and w' is the word on the tape from the left of the read-write head. Due to the conditions imposed on the productions, our Turing machine is deterministic and its behavior depends only on the present state and the scanned letter. (This is the reason that productions (2) and (4) are present for every c.) Moreover, the addition of a new blank square B at the end of the tape is taken care of, the end of the tape being marked by #. A slight difference with respect to the intuitive explanations given before the definition is the fact that our formal Turing machine either only overprints or moves the read-write head during one time instant, rather than doing both of the operations at the same time instant. Of course, this is not essential because the two operations can be performed one after another if an auxiliary state is added.

Definition 4.1 considers Turing machines only as acceptors. We still want to define output-producing devices. In particular, we want to consider the computation of values of functions (of possibly several variables).

More specifically, we consider functions

$$f(x_1, \ldots, x_k)$$

with $k \geq 1$ variables ranging over Σ_T^* for some alphabet Σ_T and with function values also in Σ_T^*. For instance, functions mapping some Cartesian power of the set N_0 of nonnegative integers into N_0 can be expressed in this fashion. This follows because if $\Sigma_T = \{a_1, \ldots, a_n\}$, then the words in Σ_T^* can be identified with the elements of N_0, represented n-adically (see Example 3.3). For instance, if $\Sigma_T = \{a_1, a_2\}$, we may identify a_1 and a_2 with the digits 1 and 2, obtaining the dyadic representation, where the empty word represents the integer 0.

We consider also **partial** functions $f(x_1, \ldots, x_k)$, meaning that the function value need not be defined for all argument values. Thus, a **total** function $f(x_1, \ldots, x_k)$ maps $(\Sigma_T^*)^k$ into Σ_T^*, whereas a partial function $f(x_1, \ldots, x_k)$ maps only some subset of $(\Sigma_T^*)^k$ into Σ_T^*.

Definition 4.2. *Consider a Turing machine TM as introduced in Definition 4.1. A partial function f from $(\Sigma_T^*)^k$ into Σ_T^* is **computed by TM** iff for all $x_1, \ldots, x_k$ in Σ_T^*, conditions (i) and (ii) are satisfied.*

(i) Assume that $f(x_1, \ldots, x_k)$ is defined. Consider a morphism h mapping all letters of Σ_T and all letters of $\Sigma_Q - \{q_F\}$ into themselves and mapping all remaining letters of Σ into λ. Then

$$\#q_0x_1Bx_2B\cdots Bx_k\# \Rightarrow^* w \quad with \quad h(w) = f(x_1, \ldots, x_k).$$

(ii) Assume that $f(x_1, \ldots, x_k)$ is undefined. Then TM loops with $x_1Bx_2B\cdots Bx_k$.

*A partial function f is called **partial recursive** iff f is computed by some Turing machine. If, in addition, f is total (that is, f is defined for all argument values from $(\Sigma_T^*)^k$), then f is called **recursive**.* □

Thus, the argument is given by using the blanks B as separators between the different argument places. In the final contents of the tape, all symbols not belonging to Σ_T are ignored. Observe that if the final state q_F

is not reached during the computation, then the h-value of the end result can never belong to Σ_T^* because h maps all the nonfinal states into themselves. Hence, the final state must be reached for all inputs for which the function is defined.

We sometimes use the term **total recursive** (rather than recursive) to emphasize that the function considered is both total and computable by a Turing machine. Sometimes, it is convenient to consider only positive (instead of nonnegative) values for variables. In this case, the term *total* means that the function is defined for all positive values of the variables. No confusion should arise because it is clear from the context which case is meant.

As already discussed in the introduction, the dilemma of diagonalization is avoided by letting the functions considered be partial. Observe that when a Turing machine is started with some input $x_1 B \cdots B x_k$, we have no a priori knowledge concerning whether or not the computation will ever halt—that is, whether or not $f(x_1, \ldots, x_k)$ is defined. Indeed, if this were not the case—if we had an effective bound on the length of the computation (depending possibly on TM and its input)—we could easily reconstruct the diagonalization dilemma and obtain an intuitively computable partial function that is not computable by any Turing machine. These matters are discussed in more detail in Section 4.3 in connection with the halting problem.

We conclude this section with an example related to a question of general philosophical interest.

It is sometimes claimed that if a function (for example, from natural numbers into natural numbers) has been defined precisely, then it can also be computed by a computer. The following example shows that this is not necessarily the case. The example is based on a special type of diagonalization.

Example 4.1. Consider Turing machines with

$$\Sigma_O = \{\#, B, 1\}, \qquad \Sigma_T = \{1\}, \qquad \Sigma_Q = \{q_0, q_1, \ldots, q_k\}$$

for some $k \geq 1$. Denote by $C(k)$ the collection of all such machines. Thus, TM being in $C(k)$ means that the operating alphabet contains only one "essential" letter 1, and that the state alphabet consists of k elements in addition to the initial state, q_0. Clearly $C(k)$ is finite for every k.

We write f_{TM} to denote the partial function of one variable computed by a Turing machine TM. Consider now the following function g mapping the set of natural numbers into itself:

$$g(k) = \max \{i \mid f_{TM}(1) = 1^i \text{ for some } TM \text{ in } C(k)\}.$$

Thus, $g(k)$ equals the greatest number of 1s any Turing machine in $C(k)$ can have on its tape after a halting computation starting from a single 1 on its tape (that is, after having $\#q_0 1\#$ as the first word in the rewriting process). Since $C(k)$ is finite and since the rewriting process is completely deterministic for every TM in $C(k)$ (that is, when started with $\#q_0 1\#$ the machine TM either halts with some number of 1s on its tape or else loops), we conclude that $g(k)$ is well defined. It is also obvious that g is a total function because in every $C(k)$ there is a Turing machine doing nothing—that is, halting immediately with the input 1. Consequently, g is indeed a function mapping the set of natural numbers into itself. There can also be no question about the fact that g is precisely defined. The reader is encouraged to compute the value of $g(k)$ for a couple of small values of k. However, a warning is in order: Very soon the number of cases to be studied becomes hopeless. (See Exercise 4.1.)

However, g is *not* recursive and hence cannot be computed by any Turing machine. We prove this indirectly. Assume that g actually is recursive. Therefore, the function $g'(x) = g(2x) + 1$ is also recursive. Indeed, this is an immediate consequence of Church's thesis, since there is an algorithm for computing the values of g'. Moreover, we may assume without loss of generality that a Turing machine $TM(g')$ computing g' is in $C(t)$ for some t. This follows because our restriction of the operating alphabet involves no loss of generality if we do not simultaneously impose an upper bound on the number of states: An arbitrary number of tape symbols can be encoded in terms of 1 and B if sufficiently many states are available for the decoding. (This is not true if only machines with, say, 10 states are considered.) We can also readily go from one representation (unary, dyadic, and so on) of natural numbers into another.

Consider now the Turing machine TM_1, first changing the input 1 into 1^t, using states $q_0, \ldots, q_{t-1}$, and then calling the machine $TM(g')$, whose states have been renamed as q_t (initial), $q_{t+1}, \ldots, q_{2t}$. Thus, the beginning of the computation of TM_1 looks like:

$$\#q_0 1\# \Rightarrow \#1q_0 B\# \Rightarrow \#1q_1 1\# \Rightarrow \#11q_1 B\# \Rightarrow \#11q_2 1\#$$
$$\Rightarrow \#111q_2 B\# \Rightarrow *\#1^{t-1}q_{t-2} B\# \Rightarrow \#1^{t-1}q_{t-1} 1\#$$
$$\Rightarrow *\#q_{t-1} 1^t\# \Rightarrow \#q_{t-1} B1^t\# \Rightarrow \#Bq_1 1^t\#,$$

after which the routine $TM(g')$ (with modified state names) is called.

Clearly the machine TM_1 is in $C(2t)$. Hence, by the definition of the function g, the number of 1s appearing on the tape after the above computation has reached a final state should be less than or equal $g(2t)$, or else TM_1 loops with the input in question. On the other hand, by the properties of $TM(g')$, the machine TM_1 halts with this input, finally producing $g(2t) + 1$ occurrences of 1. This contradiction shows that g is not recursive.

Even the following stronger result holds: The function g grows faster than any partial recursive function. More specifically, for every partial recursive function f, there is an integer x_f (computable effectively from the Turing machine representation for f) such that whenever $x > x_f$ and $f(x)$ is defined, then $g(x) > f(x)$. The details of the proof can be obtained by modifying the argument presented above and are left to the reader. $\square$

4.2. PROGRAMMING IN MACHINE LANGUAGE, CHURCH'S THESIS, AND UNIVERSAL MACHINES

From now on we forget our formal definition of a Turing machine; it is not our intention to start proving lemmas and theorems based on the formal definition. Such a formal development could be simplified by establishing results concerning (i) subroutines and (ii) special programming techniques. Here (i) involves exhibiting particular Turing machines doing a certain job, such as multiplying two integers in unary representation. Such machines can then be used as components of more-sophisticated machines. On the other hand some examples of (ii) are the usage of several "tracks" on the tape—that is, considering the elements of the operating alphabet as n-tuples $(a_1, \ldots, a_n)$ or proving lemmas to the effect that it is no loss of generality to apply several tapes and/or several read-write heads. However, even such simplifications do not keep the formal development from being extremely tedious, as well as both time- and effort-consuming. Indeed, in all existing texts such a formal development is given up sooner or later (usually sooner!), [Da1] representing perhaps the most detailed formal development. This is no wonder because formal constructions dealing with Turing machines resemble the programming in machine language during the time of early computers. For instance, every memory location (that is, a state of a Turing machine) has a specific number attached to it, and the programming has to take care of the requirement that proper numbers are used in connection with each instruction. This is to be contrasted with the much more pleasant programming in some high-level language. For Turing machines, such a high-level language is due to Church's thesis: Whenever we have an effective (in the intuitive sense) procedure for some job, we can also construct a Turing machine for the same job.

We use Church's thesis in this chapter as a tool in almost all proofs, thus avoiding the really tedious constructions involving the formal Turing machine model. We have already discussed the various reasons for accepting Church's thesis. In spite of such reasons, the reader should be able (at least in principle) to replace every argument based on Church's thesis by an argument based on formal definitions.

In the sequel we consider an enumeration

$$TM_1, TM_2, \ldots, TM_i, \ldots \tag{6}$$

of all Turing machines. Such an enumeration can be obtained, for instance, as follows. Every Turing machine, TM, is completely determined by the word $w(TM)$ obtained by writing the productions of TM one after the other, separated by some marker symbol. (Some additional convention concerning how one recognizes the terminal letters has to be made. One way to do this is to list all the alphabets involved in $w(TM)$.) Such words are then enumerated, for example, according to length and alphabetic order. (We assume that all symbols used come from a fixed denumerable collection.) If $w(TM)$ is the ith word in this enumeration, we let TM be the ith machine.

From now on we consider some fixed enumeration, such as (6). The number i is referred to as the **index,** or the **Gödel number,** of the Turing machine TM_i. Our results are independent of the actual choice of the enumeration. The only requirement is that there are effective procedures for determining TM_i, given i, and for determining i such that $TM_i = TM$, given TM. In other words, there are effective procedures for going from the index to the machine and from the machine to the index.

For all positive integers i and k, let

$$f_i(x_1, \ldots, x_k) \tag{7}$$

be the partial recursive function of k variables computed by the Turing machine TM_i. As already pointed out, we may apply m-adic notation for any terminal alphabet and view f_i as a function mapping some subset of the Cartesian power N_0^k into N_0. This is always done when, for instance, arithmetical operations are considered. The number i is also referred to as the **index of the function** in (7).

Observe that every partial recursive function $f(x_1, \ldots, x_k)$ has infinitely many indices i. This follows because to every Turing machine computing f, we may add another "dummy" state, which is never reachable during a computation.

The next theorem is usually called **Kleene's s-m-n theorem.** The proof makes use of Church's thesis in the following (somewhat extended) sense.

Consider argument (i) presented earlier in support of Church's thesis. Indeed, somewhat more can be said: All formalizations of effective procedures have turned out to be equivalent in the stronger sense that one can go *uniformly* from one formalization to another. What this means is that if (6) is compared with another listing

$$P_1, P_2, \ldots, P_i, \ldots$$

of effective procedures (in some formalization possibly different from Turing machines), then there are recursive functions g and h such that P_i and $TM_{g(i)}$, as well as TM_i and $P_{h(i)}$, are equivalent (that is, the partial recursive functions computed by them coincide). Of course, it is assumed that in the P_i-formalization, transitions from i to P_i, and vice versa, are also effective.

Let m and n be greater than or equal 1 and consider the partial recursive functions

$$f_i(y_1, \ldots, y_m, z_1, \ldots, z_n), \tag{8}$$

for $i = 1, 2, \ldots$. Clearly, every partial recursive function $g(y_1, \ldots, y_m)$ of m variables can be expressed as one of the functions in (8). For instance, we may choose an appropriate function from (8) not depending at all on the last n variables or choose an i such that

$$f_i(y_1, \ldots, y_m, 0, \ldots, 0) = g(y_1, \ldots, y_m), \tag{9}$$

for all $y_1, \ldots, y_m$. (More explicitly, this means that for every m-tuple $(y_1, \ldots, y_m)$, either both sides of (9) are undefined, or else both sides of (9) are defined and assume the same value.) On the other hand, any choice of constants for the z's in (8) gives a partial recursive function of m variables.

This implies, by Church's thesis, the existence of a (total) recursive function $s(i, z_1, \ldots, z_n)$ of $n + 1$ variables such that

$$f_i(y_1, \ldots, y_m, z_1, \ldots, z_n) = f_{s(i, z_1, \ldots, z_n)}(y_1, \ldots, y_m) \tag{10}$$

for all $i, y_1, \ldots, y_m, z_1, \ldots, z_n$. Hence, we have established the following result.

Theorem 4.1. *For all positive integers m and n, there is a total recursive function s (depending on m and n) such that identity (10) holds.*

In Kleene's s-m-n theorem, as well as elsewhere in the sequel, we have a specific indexing of partial recursive functions in mind. If the indexing is changed, the functions s might change, too, but the statement of the theorem remains unaltered.

The s-m-n theorem is a powerful tool in various constructions. As an example we mention the following result.

Theorem 4.2. *There is a recursive function $g(i, j)$ such that*

$$f_i(x) + f_j(x) = f_{g(i, j)}(x) \tag{11}$$

for all i, j, x.

Proof. We show first that the left side of (11), viewed as a function $\psi(x, i, j)$ of the three variables, is partial recursive. (The definition of the sum should be obvious: It is defined for those values of x for which both summands are defined. A similar remark applies to other operations as well.)

Consider the following intuitive algorithm for computing the values of ψ. Given an argument (x, i, j), start both of the Turing machines TM_i and TM_j on the input x. Execute by turns one step of the computation of both machines: first step of TM_i (with input x), first step of TM_j, second step of TM_i, second step of TM_j, and so forth. If and when one machine halts, continue the computation of the other machine. If and when both machines halt, output the sum of their outputs. (The execution of the computation steps of the two machines by turns is not actually necessary in this case, but it will be very important many times in the sequel.)

By Church's thesis, $\psi(x, i, j)$ is partial recursive and, consequently, has an index i_0:

$$f_i(x) + f_j(x) = \psi(x, i, j) = f_{i_0}(x, i, j).$$

By Theorem 4.1 (choosing $m = 1$, $n = 2$), we conclude the existence of a recursive function s of three variables such that

$$f_{i_0}(x, i, j) = f_{s(i_0, i, j)}(x).$$

The claim (11) now follows by choosing $g(i, j) = s(i_0, i, j)$. Clearly, g is recursive. $\qquad\square$

Theorem 4.2 shows that the sum of two partial recursive functions (viewed as a function of one variable—two different variables can be handled similarly) is again partial recursive. In fact, the result is stronger: The index of the sum is obtainable in a **uniform effective** way from the indices of the given functions. This follows because, for all i and j, the sum function has the index $g(i, j)$.

Similar results can be established by Theorem 4.1 for operations other than sum: product, composition, and so on.

Essentially, Theorem 4.1 gives a method of "lowering" the position of some variables from the argument to the index. The next theorem establishes a reverse procedure for making the position of some variables "higher."

Theorem 4.3. *For every $m \geq 1$, there exists a u such that*

$$f_i(y_1, \ldots, y_m) = f_u(i, y_1, \ldots, y_m). \tag{12}$$

Proof. We again view $f_i(y_1, \ldots, y_m)$ as a function $\psi(i, y_1, \ldots, y_m)$ of $m + 1$ variables. Given an argument $(i, y_1, \ldots, y_m)$, an obvious intuitive algorithm to compute the value $\psi(i, y_1, \ldots, y_m)$ is the following. Start the Turing machine TM_i with the argument $(y_1, \ldots, y_m)$. If and when the computation halts, output the final result of the computation.

By Church's thesis, ψ is a partial recursive function and, hence, possesses an index u. Consequently, (12) holds true. $\qquad\square$

The Turing machine TM_u is customarily referred to as the **universal Turing machine** (for partial recursive functions of m variables). Indeed, this terminology is quite natural. Whenever TM_u receives as its input the index i of a particular Turing machine TM_i followed by some m-tuple $(y_1, \ldots, y_m)$, then TM_u behaves exactly as TM_i would behave given the input $(y_1, \ldots, y_m)$.

4.3. RECURSION THEOREM AND BASIC UNDECIDABILITY RESULTS

By Church's thesis the decidability of a particular problem—that is, the existence of an algorithm for solving the problem—can be expressed in the following equivalent and precise fashion.

Consider the membership problem for a given language L, which is the problem of deciding for an arbitrary word w whether $w \in L$. Clearly, by

Proof. For each integer $a \geq 1$, we define a partial function τ_a of one variable as follows. Given an argument x, we first apply TM_a to the input a. If TM_a halts with a, producing the output b, then we apply TM_b to the input x. The output, if any, is the value $\tau_a(x)$. To summarize:

$$\tau_a(x) = \begin{cases} f_{f_a(a)}(x) & \text{if } f_a(a)\downarrow \\ \uparrow & \text{if } f_a(a)\uparrow. \end{cases} \tag{18}$$

Of course $\tau_a(x) = \uparrow$ means here that $\tau_a(x)$ is undefined. Observe that $\tau_a(x)$ may be undefined even though $f_a(a)$ is defined.

By Church's thesis, every function τ_a is partial recursive and, hence, possesses an index depending on a—for instance, $h(a)$. Since we have also given an algorithm for computing $h(a)$ given a, another application of Church's thesis tells us that h is recursive. Consequently, (18) holds with $\tau_a(x)$ replaced by $f_{h(a)}(x)$. (A more explicit version of the above argument results if one first views τ as a function of two variables a and x and then applies the *s-m-n* theorem.)

Since g and h are both recursive, so is their composition, gh. Let i be an index for gh. Since gh is total and $gh = f_i$, we conclude that $f_i(i)\downarrow$. Hence, by (18), with $\tau_a(x)$ replaced by $f_{h(a)}(x)$, we obtain

$$f_{h(i)} = f_{f_i(i)} = f_{g(h(i))}.$$

Consequently, (17) holds with $n = h(i)$. $\qquad\qquad\square$

Theorem 4.6 is the basic version of the recursion theorem. A more sophisticated version is needed in Section 4.5.

At this point, we could continue the list of undecidability results begun in Theorems 4.4 and 4.5. For instance, each of Problems (i)-(vi) is undecidable:

   (i) Is a given partial recursive function total? (In fact, if this problem were decidable, we could apply diagonalization to total functions alone.)

  (ii) Is a given partial recursive function constant?

 (iii) Is the language accepted by a given Turing machine empty?

 (iv) Is the language accepted by a given Turing machine infinite?

  (v) Do two given Turing machines accept the same language?

 (vi) Do the ranges of two given partial recursive functions coincide?

The reader is encouraged to establish the undecidability of (i)–(vi), either by a direct or by an indirect argument. The list (i)–(vi) is by no means exhaustive—it can be continued indefinitely by problems formulated either in terms of partial recursive functions or in terms of formal languages. In many cases, one can also translate one formulation into the other. For instance, Theorem 4.4 can be interpreted as the undecidability of the membership problem for languages accepted by Turing machines. (In fact, the family of these languages equals the family of recursively enumerable languages studied in Chapter 2.)

Rather than proving special undecidability results such as (i)–(vi), we shall now establish a general result to the effect that every nontrivial property of partial recursive functions is undecidable. (The result can also be formulated in terms of languages rather than functions.) A property is **nontrivial** if it is possessed by at least one partial recursive function (of one variable) but not by all of them. Obviously Problems (i)–(vi) correspond to nontrivial properties. The next theorem is known as **Rice's theorem.**

Theorem 4.7.   *Let F be a nonempty proper subset of the set of partial recursive functions of one variable. Then the characteristic function of the set*

$$S_F = \{i \mid f_i \in F\}$$

is not recursive.

Proof.   Assume the contrary: The characteristic function g of S_F is recursive. We compute the values $g(1), g(2), \ldots$ until we find numbers i_1 and i_2 such that

$$f_{i_1} \in F \quad \text{and} \quad f_{i_2} \notin F.$$

By our assumption concerning F, such numbers will eventually be found.

Since g is recursive, the function

$$g_1(x) = \begin{cases} i_2 & \text{if } f_x \in F \\ i_1 & \text{if } f_x \notin F \end{cases}$$

is also recursive. (In fact, $g_1 = hg$, where h maps 1 to i_2 and 0 to i_1.) Let n be the fixed point of g_1 according to Theorem 4.6: $f_n = f_{g_1(n)}$.

Assume that f_n is in F. Hence, $g_1(n) = i_2$. The equation $f_n = f_{g_1(n)}$ now implies that f_n is not in F.

Assume next that f_n is not in F. Hence, $g_1(n) = i_1$, showing that f_n is in F.

Hence, a contradiction arises in both cases. $\square$

Theorem 4.7 can also be established without using recursion theorem; see Exercise 4.4.

Although Theorem 4.7 shows that every nontrivial property of partial recursive functions is undecidable, it does not show that every nontrivial property of Turing machines is undecidable. For instance, each of the following nontrivial properties of *TM* is easily shown to be decidable: (i) *TM* has 3 states, (ii) started with the input 1, *TM* moves to the right all the time, and (iii) started with the input 1, *TM* visits at least one square at least 17 times. It is also easy to find a particular index i such that the halting problem of TM_i is decidable: Given an input x, we can decide whether or not TM_i halts with x.

4.4.   RECURSIVE AND RECURSIVELY ENUMERABLE SETS AND LANGUAGES

We begin with a result analogous to Theorems 2.4, 2.11, and 2.12.

Theorem 4.8.   *A language is recursively enumerable iff it is accepted by some Turing machine.*

Proof.   Given L with $L = L(TM)$, we construct a grammar G such that $L = L(G)$ as follows. The start letter of G generates an arbitrary word appearing on the tape, surrounded by boundary markers #, and containing also one occurrence of the final state. The main part of the productions consists of the productions of TM with their two sides interchanged. Finally, the boundary markers and the state symbol are eliminated by the productions $\#q_0 \to \lambda$ and $\# \to \lambda$. The formal details are left to the reader. (Also, [Sa3] may be consulted.)

Conversely, given $L = L(G)$, we use Church's thesis to find a Turing machine TM such that $L = L(TM)$. This follows because G clearly constitutes an effective procedure for listing all words of L. (If Church's thesis is not used, then the proof of the converse part is fairly complicated, the main difficulty being due to the simulation of arbitrary rewriting by monogenic rewriting.)   □

A language L being recursively enumerable means that there is an effective procedure for listing all words of L (possibly with repetitions). We say that a language L is recursive iff there is an algorithm for deciding of a given word w whether or not w is in L. In other words, L is recursive iff the characteristic function of L is recursive.

These notions will now be defined for sets of numbers.

Definition 4.3.   *A set S of nonnegative integers is **recursively enumerable** iff S equals the domain of some partial recursive function of one variable. S is **recursive** iff the characteristic function of S is recursive.*   □

We have already pointed out that every language $L \subseteq \{a_1, \ldots, a_m\}^*$ can be viewed as a set of nonnegative integers by considering words as integers in m-adic notation. We denote by $N(L)$ the set of integers associated with L in this fashion. We assume here that $\{a_1, \ldots, a_m\} = \text{ALPH}(L)$. Of course, m-adic notation presupposes a linear order among the letters of the alphabet. If L is originally over the alphabet $\{b, c\}$, then the set $N(L)$ also depends on which of the letters b and c is chosen to be a_1. However, for our purposes, this is irrelevant because the properties of the sets $N(L)$ we are considering do not depend on the ordering of the letters.

For instance, if $L \subseteq \{a_1, a_2\}^*$ is the language denoted by the regular expression $(a_1 \cup a_2)^* a_1$, then $N(L)$ equals the set of all odd integers. If $L \subseteq \{a_1, a_2\}^*$ is the language denoted by the regular expression $(a_1)^* a_2$, then

$$N(L) = \{2^n \mid n \geq 1\}. \tag{19}$$

The following results show that the definitions of recursive enumerability and recursiveness are compatible for languages and sets of numbers. The next theorem is an immediate consequence of Theorem 4.8.

Theorem 4.9. *If a language L is recursively enumerable (resp. recursive), then $N(L)$ is also recursively enumerable (resp. recursive).*

Now consider the converse of Theorem 4.9. In general, the transition from a set of numbers to a language may yield different results depending on the choice of the base. For instance, consider the set of all positive powers of 2. We have seen (recall (19)) that in dyadic notation—that is, viewed as a language over $\{a_1, a_2\}$—this set is regular. However, it is obvious that viewed as a language over $\{a_1\}$, the same set is nonregular. For instance, it can also be shown that if $m = 3$ or $m = 5$, then the set (19) is nonregular when viewed as an m-adic set.

The notions of recursive enumerability and recursiveness are still general enough to be independent of the choice of the base m. In other words, if a set of m-adic integers is recursively enumerable (resp. recursive), then it is also recursively enumerable (resp. recursive), viewed as a set of m_1-adic integers, for any $m_1 \geq 1$. The reader should have no difficulties in establishing this fact, thus obtaining the following converse of Theorem 4.9.

Theorem 4.10. *If a set N of nonnegative integers is recursively enumerable (resp. recursive), then any language L such that $N = N(L)$ is also recursively enumerable (resp. recursive).*

Theorem 4.10 shows that the notions of recursive enumerability and recursiveness are highly invariant and, consequently, most natural. In view of Theorems 4.9 and 4.10, we shall restrict our attention to sets of numbers (rather than to languages) from now on in this section.

Let us first discuss in detail why the recursive enumerability of a set S implies that there is an effective procedure for listing the elements of S. Thus, assume S is the domain of some partial recursive function g, computed by a Turing machine TM. To list S, we consider the computation of TM for the inputs $0, 1, 2, \ldots$. Whenever the computation halts, we output the corresponding input. We still have to order the steps of this procedure in a proper fashion to avoid the situation in which we would be forever computing a particular input i for which TM loops. This can be accomplished by a "dovetailing" method, resembling the pairing function φ. Denote by $\alpha(i, j)$ the jth step in the computation of TM for the input i. Then we take the steps in the order

$$\alpha(0, 1), \alpha(0, 2), \alpha(1, 1), \alpha(0, 3), \alpha(1, 2), \alpha(2, 1), \alpha(0, 4), \ldots,$$

making sure that for all i and j, the jth step of the computation of TM for the input i eventually appears in the list (provided TM does not halt for i before the jth step).

We shall now establish some simple results showing the interconnection between recursiveness and recursive enumerability.

Theorem 4.11. *Every recursive set is recursively enumerable.*

Proof. Let S be a set whose characteristic function g is recursive. Let g be computed by the Turing machine TM. We may obviously modify TM to yield a machine TM' such that TM' loops whenever TM produces the output 0, whereas TM' behaves exactly as TM in case TM produces the output 1. □

Theorem 4.12. *A set S is recursive iff both S and its complement $\sim S$ are recursively enumerable.*

Proof. The *only if* part follows by Theorem 4.11 and by the observation that whenever S is recursive, then so is $\sim S$. Consider the *if* part and assume that both S and $\sim S$ are recursively enumerable. Hence, there are effective procedures for listing both the elements $a_1, a_2, \ldots$ of S and the elements $b_1, b_2, \ldots$ of $\sim S$, possibly with repetitions. Given an arbitrary integer x, we now go through the two lists in a zigzag fashion until we find x. This constitutes an algorithm for deciding whether x is in S or in $\sim S$. Consequently, S is recursive. □

It remains to be shown that recursively enumerable sets are more general than recursive ones.

Theorem 4.13. *There are recursively enumerable sets that are not recursive. An example is the set of self-applicable indices*

$$SA = \{i \mid f_i(i)\!\downarrow\}.$$

Proof. That SA is recursively enumerable is obvious: Given i, we just start the ith Turing machine with the input i and see what happens. In this fashion, we can express SA as the domain of a partial recursive function. On the other hand, SA being recursive contradicts Theorem 4.5. □

The following example reviews the different methods of showing undecidability.

Example 4.2. The set

$$K_7 = \{i \mid \text{domain of } f_i \text{ equals domain of } f_7\}$$

is not recursive. To prove this, assume the contrary: K_7 has a recursive characteristic function g. We now establish a contradiction by three different methods, (i)–(iii), making use of the recursion theorem, diagonalization, and the *s-m-n* theorem, respectively. We denote the domain of a function f by DOMf.

(i) Choose an index i_0 such that

$$\text{DOM} f_{i_0} \neq \text{DOM} f_7.$$

(Clearly, such an index exists!) Define

$$h(i) = \begin{cases} i_0 & \text{if } g(i) = 1 \\ 7 & \text{if } g(i) = 0. \end{cases}$$

Since h is clearly recursive, we conclude the existence of an n such that $f_n = f_{h(n)}$, by the recursion theorem. But now

$$g(n) = 1 \quad \text{implies} \quad h(n) = i_0 \quad \text{implies} \quad \text{DOM} f_{h(n)}$$
$$= \text{DOM} f_n \neq \text{DOM} f_7 \quad \text{implies} \quad g(n) = 0.$$

On the other hand,

$$g(n) = 0 \quad \text{implies} \quad h(n) = 7 \quad \text{implies} \quad \text{DOM} f_{h(n)}$$
$$= \text{DOM} f_n = \text{DOM} f_7 \quad \text{implies} \quad g(n) = 1.$$

Consequently, we have a contradiction in both cases.

(ii) We assume first that DOMf_7 is infinite. Let $a_1, a_2, \ldots$ be an enumeration of DOMf_7 without repetitions, Since g is recursive, we may also list all the (infinitely many) indices in the set $K_7 : i_1, i_2, \ldots$. Define

$$h(x) = \begin{cases} f_{i_m}(a_m) + 1 & \text{for } x = a_m \\ \uparrow & \text{for } x \notin \text{DOM} f_7. \end{cases}$$

Clearly, h is partial recursive and DOM$h = $ DOMf_7. However, the index of h is not in K_7, a contradiction.

The case where DOMf_7 is finite is somewhat more complicated to handle by diagonalization. We begin by determining DOMf_7. This can be done effectively as follows. When a subset A of DOMf_7 has been enumerated (possibly A is empty), we construct a function f_A whose domain equals A and apply the function g to decide whether or not $A = $ DOM$f_A = $ DOMf_7. This allows us to check when our enumeration of DOMf_7 has exhausted the whole DOMf_7.

We now enumerate the (infinitely many) elements of $\sim K_7 : j_1, j_2, \dots$. Clearly, the indices of all functions with an infinite domain belong to this list. A new list of indices $k_1, k_2, \dots$ is then constructed as follows.

We first modify the function f_{j_i} in such a way that the new function will assume the value 1 for arguments in DOMf_7 and the value $f_{j_i}(x)$ for other arguments x. Using g, we check whether or not the domain of the modified function equals DOMf_7. If it does, we proceed in the same way with index j_2, and so forth, until we find inequality. Let k_1 be the index of the function found. We may then determine a number x_1 in the difference DOMf_{k_1} − DOMf_7.

We continue with the j-indices as before, but now we make the additional modification that the functions diverge for the argument value x_1. In this way, we find the index k_2, the corresponding number x_2, and so forth. The function h is now defined by

$$h(x) = \begin{cases} f_{k_i}(x_i) + 1 & \text{for } x = x_i) \\ \uparrow & \text{if } x \text{ is not one of the } x_i\text{'s.} \end{cases}$$

A contradiction now arises because the index of h does not belong to $\sim K_7$, although it should.

(iii) We distinguish two cases, depending whether DOMf_7 is empty or nonempty. Assume first that DOMf_7 is empty. Define

$$h(x, y) = \begin{cases} 0 & \text{if } f_x(x)\downarrow \\ \uparrow & \text{otherwise.} \end{cases}$$

By Church's thesis and the *s-m-n* theorem, there is a recursive function k such that

$$h(x, y) = f_{k(x)}(y).$$

But now the recursive function gk is apparently characteristic for $\sim SA$, a contradiction.

Assume, secondly, that DOMf_7 is nonempty. Define

$$h(x, y) = \begin{cases} f_7(y) & \text{if } f_x(x)\downarrow \\ \uparrow & \text{otherwise.} \end{cases}$$

Again, there is a recursive function k such that $h(x, y) = f_{k(x)}(y)$. Now the recursive function gk is characteristic for the set SA, which is also a contradiction. $\square$

Let S be a recursively enumerable set. Consequently, S equals the domain of some partial recursive function $f(x)$. If i is an index for f, we say that i is also an **index for the set** S.

Example 4.3 deals with fixed-point properties of indices for recursively enumerable sets. Example 4.4 discusses a particularly complicated set.

Example 4.3. Suppose we want to find an index i such that $S_i = \{i\}$. In other words, the ith set (according to our indexing) consists of i alone. At a first glance, we might think that this is not always possible. In particular, we could have chosen the indexing in such a way that no fixed-point situations $S_i = \{i\}$ arise. (Recall that the choice of the indexing was quite arbitrary, apart from the conditions of effectiveness.) However, such a "vicious" choice of the indexing is not possible: A fixed point can be found for any indexing. An intuitive reason for this is that enough leeway is always provided by the fact that every recursively enumerable set has infinitely many indices. The formal argument uses recursion theorem in the following fashion.

Given a number i, we may easily construct a Turing machine $TM(i)$ such that the domain of the partial recursive function of one variable, computed by $TM(i)$, equals i. Indeed, $TM(i)$ just checks whether or not the input is i. If it is, it halts. Otherwise, it loops. The machine $TM(i)$ can be constructed in a uniform effective way. By Church's thesis, there is a recursive function $g(x)$ such that the domain of the function $f_{g(i)}$ equals i. By Theorem 4.6, there is a natural number n such that

$$f_n = f_{g(n)}.$$

Since the functions coincide, their domains must also coincide. Consequently, $S_n = S_{g(n)}$. Since $S_{g(n)} = \{n\}$, we conclude that $S_n = \{n\}$.

The same method can be applied in a number of similar constructions. For instance, we can find an i such that $S_i = \{i, i^2\}$.

Example 4.4. The set SA of Theorem 4.13 is an example of a recursively enumerable set whose complement is not recursively enumerable. If $\sim SA$ were recursively enumerable, then SA itself would be recursive, a contradiction. Clearly, the complement of a recursive set is recursive and, hence, recursively enumerable. We thus have examples of sets S such that (i) both S and $\sim S$ are recursively enumerable, as well as of sets S such that (ii) exactly one of the sets S and $\sim S$ is recursively enumerable. We now give an example of a set S such that (iii) neither S nor $\sim S$ is recursively enumerable. Sets (iii) are more complicated than sets (ii) which, in turn, are more complicated than sets (i) with respect to the reducibilities considered in the next section.

Consider the set

$$R = \{i \mid S_i \text{ in dyadic notation is a regular language}\}.$$

For instance, every index of the set $\{2^n \mid n \geq 1\}$ is in R. In dyadic notation, the latter set can be expressed as $(a_1)^*a_2$. The dyadic notation is by no means essential for the definition of R; we just wanted to specify in a unique fashion the language corresponding to a set of numbers.

We now prove that neither R nor $\sim R$ is recursively enumerable. Consider first the function

$$\tau(x, y) = \begin{cases} 1 & \text{if } f_x(x)\downarrow \text{ and } y \text{ is prime} \\ \uparrow & \text{in other cases.} \end{cases}$$

By Church's thesis and the s-m-n theorem, $\tau(x, y)$ can be expressed as $f_{g(x)}(y)$, where g is a recursive function. Consequently, $S_{g(x)}$ is either the set of primes or the empty set, depending on whether $f_x(x)\downarrow$ or $f_x(x)\uparrow$. Hence,

$$S_{g(x)} \quad \text{in dyadic notation is regular iff} \quad f_x(x)\uparrow. \tag{20}$$

(It is clear that the set of primes in unary notation is nonregular. The same holds true for every m-adic notation. The proof concerning the dyadic notation considered in this example is left to the reader. Of course, R can also be defined in terms of unary notation, in which case its nonregularity is obvious.)

It is a consequence of (20) that R is not recursively enumerable. Assume the contrary. We can then decide whether $f_x(x)\downarrow$ or $f_x(x)\uparrow$ by the following algorithm. At odd steps of the algorithm, we perform one step of the computation of the Turing machine TM_x started with the input x. At even steps, we generate one element of R (possible because R is recursively enumerable!) and check whether it equals $g(x)$. This algorithm terminates because if $f_x(x)\uparrow$, then $g(x)$ is eventually generated. This contradiction with Theorem 4.5 shows that R is not recursively enumerable. (We have, in fact, also shown that $\sim SA$ is recursively enumerable, contradicting Theorem 4.13.)

The argument is slightly modified to show that $\sim R$ is not recursively enumerable. Consider now the function

$$\tau_1(x, y) = \begin{cases} 1 & \text{if } TM_x \text{ with input } x \text{ does not halt in } y \text{ or fewer} \\ & \text{steps and } y \text{ is prime,} \\ \uparrow & \text{otherwise.} \end{cases}$$

Again, $\tau_1(x, y)$ can be expressed as $f_{h(x)}(y)$ for some recursive function h. In this case, $S_{h(x)}$ equals the set of primes if $f_x(x)\uparrow$, and $S_{h(x)}$ equals some finite set if $f_x(x)\downarrow$. Consequently, (20) takes now the form:

$$S_{h(x)} \quad \text{in dyadic notation is nonregular iff} \quad f_x(x)\uparrow.$$

The rest of the argument is as before. □

The next theorem gives an alternative characterization of recursively enumerable sets.

Theorem 4.14. *A set is recursively enumerable iff it equals the range of some recursive function or else is empty.*

Proof. Consider the *only if* part. Let S be a nonempty recursively enumerable set. List the elements $a_1, a_2, \ldots$ of S by the method M indicated before Theorem 4.11. Define a function g as follows: $g(1) = a_1$. Assume that $g(i)$ has already been defined for $i \geq 1$. Then $g(i + 1) = a$ if the $(i + 1)$st step of M produces the element a to the list. Otherwise, $g(i + 1) = g(i)$. Clearly, g is a recursive function and, moreover, S equals the range of g.

Consider the *if* part. Obviously, the empty set is recursively enumerable. Assume that S equals the range of some recursive function g. An effective procedure for listing S consists of computing all the values of g in succession. Now consider the Turing machine TM, which—for an input x—checks whether x appears in the list. If and when it does, TM halts. Clearly the domain of the partial recursive function computed by TM equals S.   □

Recursively enumerable and recursive sets possess various closure properties. For instance, the union and intersection of two recursively enumerable (resp. recursive) sets is again recursively enumerable (resp. recursive). The closure is effective; for any two sets (of either type), their union (of the same type) can effectively be constructed. Moreover, the closure is uniform effective: There is a recursive function g such that $S_i \cup S_j = S_{g(i, j)}$ holds for all indices i and j. These facts, as well as various other closure properties, can be established by Church's thesis similar to, for instance, Theorem 4.2.

The situation is trickier in regard to complementation. Recursively enumerable sets are not closed under complementation, as shown in Theorem 4.13. It is an immediate consequence of the definition that the complement of a recursive set is again recursive. However, our final theorem in this section shows that the closure under complementation cannot be uniform effective.

Theorem 4.15.   *There is no partial recursive function τ such that for all indices i, S_i being recursive implies that $\tau(i)\downarrow$ and $S_{\tau(i)} = \sim S_i$.*

Proof.   First consider the set SA defined in Theorem 4.13. SA is recursively enumerable and, hence, effectively listable. Given an index i, we define a set as follows. We check whether i appears in SA. If and when it does, the set defined equals the whole set N of positive integers. Otherwise, we just keep trying. Hence, there is a recursive function g with the property

$$ S_{g(i)} = \begin{cases} N & \text{if } i \in SA \\ \varnothing & \text{if } i \notin SA. \end{cases} $$

Consequently, for every i, $S_{g(i)}$ is recursive. (Observe that we have described the construction of the function g in an informal manner. The reader might want to follow the more formal procedure: Start with a function of two variables and then apply Church's thesis and the *s-m-n* theorem.)

Now assume the contrary: There is a partial recursive function τ with the properties stated in Theorem 4.15. Then $g_1 = \tau g$ is a total recursive function such that, for all i,

$$S_{g_1(i)} = \begin{cases} \varnothing & \text{if } i \in SA \\ N & \text{if } i \notin SA. \end{cases}$$

Therefore,

$$\sim SA = \{i \mid S_{g_1(i)} \neq \varnothing\}. \tag{21}$$

Equation (21) gives now a possibility for listing $\sim SA$. Consequently, $\sim SA$ would be recursively enumerable, which would contradict Theorem 4.13.

The listing can be accomplished in the following fashion. For every triple (i, j, k), perform the kth step of the computation of $TM_{g_1(i)}$ for the input j. Whenever the computation halts, add i to the output list. Moreover, dovetailing is applied to order the triples so that we are not stuck forever with one particular machine computing one particular input (for which the machine might loop). For instance, we may choose in some order all the finitely many triples for which the sum $i + j + k$ is a constant n and then consider increasing values of n. $\qquad\square$

4.5. REDUCIBILITIES AND CREATIVE SETS

We have discussed various undecidable problems. In terms of the sets S_i, undecidability means that the corresponding set (by a suitable encoding) is not recursive. For instance, if a set is recursively enumerable but not recursive (like the set SA), then the corresponding problem (self-applicability problem of Turing machines) is undecidable. Thus, on the basis of decidability, we get a classification of sets into recursive and nonrecursive ones.

There are various natural ways of obtaining finer classifications. In Chapter 6 we discuss computational complexity—that is, how difficult the solution of a decidable problem is. In this way, we obtain a classification of recursive sets.

In this section, we proceed in the other direction and consider classifications of nonrecursive sets. The idea is basically the following. We may classify a problem P' to be at least as difficult as a problem P if an eventual algorithm for solving P' yields an algorithm for solving P. We may visualize this as having an "oracle" capable of solving P'. With the aid of the oracle, we can solve also P.

If we are dealing with undecidable problems and if the conclusion is not valid in the other direction (that is, an oracle for P cannot be used to solve P'), then we may view P' as being more undecidable than P.

Before formal definitions, we give more intuitive background and an example. For sets S and S' of nonnegative integers (not necessarily recursively enumerable), we write $S \leq S'$ iff any algorithm for solving the membership problem of S' yields an algorithm for solving the membership problem of S. If both $S \leq S'$ and $S' \leq S$, we write $S \equiv S'$. The relation $\leq$ is reflexive and transitive, so the relation $\equiv$ is obviously an equivalence relation. The equivalence classes are referred to as **degrees.** Thus, if two sets belong to the same degree, it can be interpreted to mean that they are equally undecidable.

If S' is recursively enumerable and for all recursively enumerable sets S we have $S \leq S'$, then S' is referred to as **complete.** Hence, if S' is complete, it represents the most-difficult problem among recursively enumerable problems.

Observe that if S is recursive and A is arbitrary, then $S \leq A$. This follows because, independently of A, we have an algorithm for solving the membership problem of S. (This is true for the intuitive notion but not for all formalizations given below. In particular, it is not true for m-reducibility.)

Before making the above notions precise, we still consider an example.

Example 4.5.   Consider the set R introduced in Example 4.4, as well as the set SA. Example 4.4 showed how the recursive enumerability of R gives a method of solving the membership problem of SA. Since the existence of an algorithm for solving the membership problem of R clearly implies that R is recursively enumerable, we conclude that $SA \leq R$. On the other hand, there is no way of using an algorithm for the membership problem of SA to settle the membership problem of R. Hence, we do not have $R \leq SA$. For a proof of this fact, however, the relation $\leq$ has to be formalized. Indeed, if the existence of an algorithm for SA can be used in every possible way, then an argument based on reductio ad absurdum proves anything!   □

We are now ready for the basic definition.

*Definition 4.4.   Consider sets S and S' consisting of nonnegative integers. S is **m-reducible** (resp. **1-reducible**) to S'—in symbols, $S \leq {}_mS'$ (resp. $S \leq {}_1S'$)—iff there is a recursive function (resp. a one-to-one recursive function) g such that for every i, i is in S exactly in case $g(i)$ is in S'. If this condition is satisfied, we also say that S is m-reducible (resp. 1-reducible) to S' via g.*   □

The condition in the above definition can also be expressed in the following equivalent forms: (i) $g(S) \subseteq S'$ and $g(\sim S) \subseteq \sim S'$, or (ii) $S = g^{-1}(S')$. However, the condition $S' = g(S)$ is *not* equivalent.

It is obvious by the definition that

$$S \leq_1 S' \quad \text{implies} \quad S \leq_m S'.$$

The converse does not hold, as will be seen below. It is also clear that $\leq_1$ and $\leq_m$ are reflexive and transitive relations. If they hold between two sets, they hold between their complements as well.

From the mathematical point of view, the relations $\leq_1$ and $\leq_m$ are the most natural reducibilities and are customarily referred to as **strong reducibilities.** We shall focus the attention on them. **Weak reducibilities** are briefly discussed at the end of this section.

Example 4.6. Consider the set

$$HALT = \{i \mid i = \varphi(x, y) \text{ for some } x \text{ and } y \text{ such that } x \in S_y\}.$$

Here φ is the pairing function discussed in Section 4.3. The set $HALT$ represents the halting problem; solving the membership in $HALT$ is equivalent to solving the halting problem.

We show first that every recursively enumerable set is 1-reducible (and consequently also m-reducible) to $HALT$. Let S be an arbitrary recursively enumerable set and let i_0 be an index for S. Then, for every x,

$$x \in S \quad \text{iff} \quad \varphi(x, i_0) \in HALT.$$

Consequently, $S \leq_1 HALT$ via the one-to-one recursive function $\varphi(x, i_0)$.

Since the set SA (defined in Theorem 4.13) is recursively enumerable, we obtain $SA \leq_1 HALT$. We now establish the 1-reducibility in the other direction. It is easy to see that $HALT \leq_m SA$. Consider a recursive function h such that

$$f_{h(x)}(y) = \begin{cases} 1 & \text{if } f_{\psi_2(x)}(\psi_1(x))\downarrow \\ \uparrow & \text{otherwise.} \end{cases}$$

(Here ψ_1 and ψ_2 are the inverses of the pairing function φ, as defined in Section 4.3. The existence of a recursive h follows by the method that should be familiar by now: Church's thesis and the s-m-n theorem.) Clearly, for every x, x is in $HALT$ iff $h(x)$ is in SA. (Observe that $f_{h(x)}$ either equals the constant 1 or else diverges everywhere.) We conclude that $HALT \leq_m SA$.

We now strengthen this result to obtain $HALT \leq_1 SA$. For this purpose, the function h is replaced by a one-to-one recursive function h_1. The technique used here is customarily referred to as *padding*. The idea is that for every x, there are infinitely many effectively obtainable indices y such that $f_x = f_y$. This provides enough leeway to replace $h(x)$ by a one-to-one function.

Consider a recursive function $g(x, y)$ such that for all y, $f_{g(x, y)} = f_x$ and, whenever either $x_1 \neq x_2$ or $y_1 \neq y_2$, then $g(x_1, y_1) \neq g(x_2, y_2)$. The function g is easily constructed by the idea mentioned above. When defining g, the arguments (x, y) are ordered according to the pairing function φ.

We now define $h_1(x) = g(h(x), x)$. Obviously, h_1 is a recursive one-to-one function. By the properties of h and g,

$$x \in HALT \quad \text{iff} \quad h_1(x) \in SA$$

holds for every x. We conclude that $HALT \leq_1 SA$.   □

Let us return again to Example 4.5. It was, in fact, shown in Example 4.4 that $SA \leq_m R$. By padding, this result can be strengthened to the result $SA \leq_1 R$. To do this, the function h of Example 4.4 has to be replaced by a one-to-one recursive function. (However, be warned: Padding does not always work!)

To show that $R \leq_m SA$ does not hold (and, hence, that $R \leq_1 SA$ does not hold), we make the following simple observation. Whenever S' is recursively enumerable (resp. recursive) and either $S \leq_m S'$ or $S \leq_1 S'$ holds, then S is also recursively enumerable (resp. recursive). This follows by the definition of the relations $\leq_m$ and $\leq_1$. By the definition, an effective procedure for listing S' (resp. an algorithm for solving the membership of S') yields an effective procedure for listing S (resp. an algorithm for solving the membership of S).

In the same way, we can conclude also that $\sim SA \leq_m SA$ does not hold. Consequently, $SA \leq_m \sim SA$ does not hold. (Recall that whenever $\leq_m$ holds between two sets, it also holds between their complements.) This means that the sets SA and $\sim SA$ are incomparable with respect to the relation $\leq_m$.

Degrees and completeness can now be defined exactly as in the informal discussion at the beginning of this section.

Definition 4.5. *If both $S \leq_m S'$ and $S' \leq_m S$ (resp. $S \leq_1 S'$ and $S' \leq_1 S$), we write $S \equiv_m S'$ (resp. $S \equiv_1 S'$). The equivalence classes of the (equivalence) relation $\equiv_m$ (resp. $\equiv_1$) are referred to as **m-degrees** (resp. **1-degrees**). A recursively enumerable set S' is **m-complete** (resp. **1-complete**) iff, for all recursively enumerable sets S, we have $S \leq_m S'$ (resp. $S \leq_1 S'$).* □

It is clear by the definition that 1-completeness implies m-completeness. Our major result in the sequel also shows that the converse implication holds. This is to be contrasted with the fact that, in general, $S \leq_m S'$ does not imply $S \leq_1 S'$. A trivial example is that $\{1, 2\} \leq_m \{1\}$ holds, but $\{1, 2\} \leq_1 \{1\}$ does not hold.

It was shown in Example 4.6 that the set $HALT$ (which clearly is recursively enumerable) is 1-complete and that

$$SA \equiv_1 HALT.$$

Clearly, for all sets S and S', if $S \equiv_1 S'$ and S' is 1-complete, then S is also 1-complete. Consequently, SA is also 1-complete. From this we infer that both $HALT$ and SA are m-complete.

We now introduce two further notions, those of productive and creative sets. We use these notions only for the characterization of m-completeness and 1-completeness. For a more-detailed discussion, the reader is referred to [Ro]. In particular, there are many interconnections with logic. For instance, the term *creative* was originally introduced by Post to reflect the fact that the set of provable formulas in a strong-enough formal system forms a creative set. According to Post, this is an indication of the creative nature of mathematics.

Before giving a formal definition, let us consider an example. We know that the set $\sim SA$ is not recursively enumerable. Suppose, however, that somebody claims that $\sim SA$ is recursively enumerable and possesses an index i. Then we can show that he or she is wrong simply by pointing out that according to the definition of SA, the index i itself must belong to the difference $\sim SA - S_i$. (More specifically, the inclusion $S_i \subseteq \sim SA$ implies that whenever x is in S_i, then $f_x(x)\uparrow$. Hence $f_i(i)\uparrow$ because $f_i(i)\downarrow$ would imply that i is in S_i, which in turn implies that $f_i(i)\uparrow$. Consequently, i is in $\sim SA$. It cannot be in S_i because, otherwise, $f_i(i)\downarrow$, which would show that i is in SA.)

Consequently, whenever some recursively enumerable set S_j is contained in $\sim SA$, we can effectively find a number in the difference $\sim SA - S_j$. (In fact, the number is simply j in this case.) This is the idea behind the definition of productive sets.

Definition 4.6. *A set S consisting of nonnegative integers is termed* **productive** *iff there is a partial recursive function f such that, for all i, the inclusion $S_i \subseteq S$ implies that $f(i) \in S - S_i$. (Hence, if $S_i \subseteq S$, then $f(i)\downarrow$.) We also say that S is productive via f. A recursively enumerable set is termed* **creative** *iff its complement is productive.* □

Our next theorem is almost an immediate consequence of the definitions.

Theorem 4.16. *No productive set is recursively enumerable. No creative set is recursive. If S is productive and $S \leq_m S'$, then S' is also productive. Every m-complete set is creative.*

We showed above that the set $\sim SA$ is productive. Consequently, by Theorem 4.16, both of the sets SA and $HALT$ are creative. We consider another example.

Example 4.7. We claim that the set

$$DIFF = \{i \mid i = \varphi(x, y) \text{ for some } x \text{ and } y \text{ such that } f_x \neq f_y\}$$

is productive. By our familiar technique, we first find a recursive function h such that

$$f_{h(x)}(y) = \begin{cases} 1 & \text{if } f_x(x)\downarrow \\ \uparrow & \text{otherwise.} \end{cases}$$

Let i_0 be an index for the constant function 1. Define $g(x) = \varphi(h(x), i_0)$. Clearly g is recursive. But now x is in SA iff $g(x)$ is in $\sim DIFF$, showing that $SA \leq_m \sim DIFF$. Hence, it is also true that $\sim SA \leq_m DIFF$. Since $\sim SA$ is productive, Theorem 4.16 shows that $DIFF$ is productive. $\square$

We have exhibited two disjoint subclasses of the class of recursively enumerable sets: recursive sets and creative sets. The two subclasses by no means cover the whole class of recursively enumerable sets. There are sets in the latter class not belonging to either one of the subclasses.

We shall now prove that the class of creative sets equals the class of m-complete—as well as the class of 1-complete—sets, showing also that the latter two classes are equal. We first need a stronger version of the recursion theorem.

Theorem 4.17. *For every recursive function $g(x, y)$, there is a one-to-one recursive function n_g such that, for every x,*

$$f_{n_g(x)} = f_{g(n_g(x), x)}.$$

The proof of Theorem 4.17 is left to the reader. The proof follows the line of the original recursion theorem (Theorem 4.6). The additional observations needed are (i) the construction of the fixed-point is uniform in x and (ii) padding can be used to make n_g one-to-one. Observe also that Theorem 4.6 can be obtained as a special case of Theorem 4.17. For instance, we may choose the function $g(x, y)$ in such a way that $g(x, 1) = g_1(x)$, where g_1 is the given function of one variable and consider the value $n_g(1)$. The following example gives a further application of Theorem 4.17.

Example 4.8.   We claim that there is a recursive function h such that for all x,

$$S_{h(x)} = \{h(x)\} \cup S_x.$$

Indeed, by Church's thesis, there is a recursive function $g(x, y)$ such that

$$S_{g(x, y)} = \{x\} \cup S_y$$

for all x and y. By Theorem 4.17, there is a recursive function $n_g = h$ such that for all x,

$$S_{h(x)} = S_{g(h(x), x)} = \{h(x)\} \cup S_x.$$ $\square$

Theorem 4.18.   *If a set S is productive, it is productive via a total recursive function.*

Proof.   Assume S is productive via the partial recursive function f. We construct, by our familiar technique consisting of Church's thesis and the s-m-n theorem, a recursive function g such that for all x and y,

$$S_{g(x,y)} = \begin{cases} S_y & \text{if } f(x)\downarrow \\ \varnothing & \text{if } f(x)\uparrow. \end{cases}$$

By Theorem 4.17, there is a recursive function $n_g = n$ such that for all y,

$$S_{n(y)} = S_{g(n(y),y)} = \begin{cases} S_y & \text{if } fn(y)\downarrow \\ \varnothing & \text{if } fn(y)\uparrow. \end{cases} \tag{22}$$

(We do not here need the fact that n can be chosen one-to-one.)
   We claim that the function fn is total. Assume the contrary: There is a number y with the property $fn(y)\uparrow$. By (22), $S_{n(y)}$ is empty and, hence, is contained in S. But since S is productive via f, we conclude that $fn(y)\downarrow$, which is a contradiction. Consequently, fn is total.
   By (22), this implies that $S_{n(y)} = S_y$, for all y. To show that S is productive via fn, we assume that $S_y \subseteq S$. Hence, $S_{n(y)} \subseteq S$. Since S is productive via f, this implies that

$$fn(y) \in S - S_{n(y)}.$$

Using the equality $S_{n(y)} = S_y$ once more, we infer that $fn(y) \in S - S_y$. This shows that S is productive via the total function fn. $\square$

Theorem 4.19.   *If a set S is productive, it is productive via a total recursive one-to-one function.*

Proof.   By the preceding theorem, we may assume that S is productive via a total recursive function f. Let $h(x)$ be a recursive function such that for all x,

$$S_{h(x)} = S_x \cup \{f(x)\}.$$

We now define a total recursive one-to-one function g such that S is productive via g.
   The starting point of the definition is: $g(1) = f(1)$. For an arbitrary $n \geq 1$, if

$$f(n + 1) \notin \{g(1), \ldots, g(n)\}, \tag{23}$$

define $g(n + 1) = f(n + 1)$. If

$$f(n + 1) \in \{g(1), \ldots, g(n)\},$$

generate values

$$fh(n + 1), fh^2(n + 1), fh^3(n + 1), \ldots \tag{24}$$

until you find an i satisfying

$$fh^i(n + 1) \notin \{g(1), \ldots, g(n)\}, \tag{25}$$

or a repetition in the sequence (24).

If such an i is found, define $g(n + 1) = fh^i(n + 1)$, with i being the smallest number satisfying (25). If a repetition occurs in the sequence in (24) before any such i is found, choose $g(n + 1)$ arbitrarily in such a way that $g(n + 1) \notin \{g(1), \ldots, g(n)\}$. (Of course, some explicit way of choosing $g(n + 1)$ should be agreed upon.) Clearly, the function g defined in this way is a total recursive one-to-one function.

We still have to show that $g(i) \in S - S_i$ whenever $S_i \subseteq S$. This is clear for $i = 1$ and also for $i = n + 1$ if (23) holds because S is productive via f.

Next, consider, the more difficult case, where

$$f(n + 1) \in \{g(1), \ldots, g(n)\}.$$

By the definition of h,

$$S_{h^i(n + 1)} = S_{h^{i-1}(n + 1)} \cup \{fh^{i - 1}(n + 1)\} \tag{26}$$

holds for all $i \geq 1$. Since S is productive via f,

$$S_{n + 1} \subseteq S \text{ implies } S_{h^i(n + 1)} \subseteq S \qquad \text{for all } i \geq 1. \tag{27}$$

Indeed, (26) reads for $i = 1$ as follows:

$$S_{h(n + 1)} = S_{n + 1} \cup \{f(n + 1)\}. \tag{28}$$

If $S_{n + 1} \subseteq S$, we must have $f(n + 1) \in S$, from which we infer by (28) that $S_{h(n + 1)} \subseteq S$. Proceeding inductively by (26), we obtain (27).

First assume that no repetition was found in the sequence (24) before an i satisfying (25) was found. If $S_{n + 1}$ is not a subset of S, there are no requirements for $g(n + 1)$ from the point of view of productivity. Hence, assume that $S_{n + 1} \subseteq S$. Then $fh^i(n + 1) \in S$, by (26) and (27), applied for the value $i + 1$. On the other hand, $fh^i(n + 1)$ cannot be in $S_{h^i(n + 1)}$ since $S_{h^i(n + 1)} \subseteq S$ and S is productive via f. By (26), $fh^i(n + 1)$ cannot be in $S_{n + 1}$ either. Hence,

$$g(n + 1) \in S - S_{n + 1},$$

as it should.

Assume, secondly, that a repetition occurs in the sequence (24). Then we cannot have $S_{n + 1} \subseteq S$ because, otherwise, $S_{h^i(n + 1)} \subseteq S$ for all i, contradicting the fact that S is productive via f. Thus, in this case there are again no requirements for $g(n + 1)$ from the point of view of productivity.

We have thus shown that the inclusion $S_{n+1} \subseteq S$ implies the condition $g(n+1) \in S - S_{n+1}$ in the more-difficult case as well, completing the proof. $\qquad\square$

Theorem 4.20.   *If a set S is creative, it is 1-complete.*

Proof.   By Theorem 4.19, we assume that $\sim S$ is productive via a one-to-one recursive function h. Consider an arbitrary recursively enumerable set A. Construct (by our familiar technique) a recursive function g such that, for all x and y,

$$S_{g(x,y)} = \begin{cases} \{h(x)\} & \text{if } y \in A \\ \varnothing & \text{if } y \notin A. \end{cases}$$

By Theorem 4.17, there is a one-to-one recursive function $n_g = n$ such that for all y,

$$S_{n(y)} = S_{g(n(y),y)} = \begin{cases} \{hn(y)\} & \text{if } y \in A. \\ \varnothing & \text{if } y \notin A. \end{cases} \tag{29}$$

Since both h and n are one-to-one, so is their composition, hn. We want to prove that A is reducible to S via hn.

Assume first that y is in A. By (29), $hn(y)$ is in $S_{n(y)}$. This implies that $S_{n(y)}$ is not contained in $\sim S$ because otherwise $hn(y)$ would not be in $S_{n(y)}$, since $\sim S$ is productive via h. Since $S_{n(y)}$ is a singleton, it must be contained in S, showing that $hn(y)$ is in S.

Assume, secondly, that y is not in A. By (29), $S_{n(y)}$ is empty, showing that $S_{n(y)}$ is contained in $\sim S$. Since $\sim S$ is productive via h, we conclude that $hn(y)$ is in $\sim S$ and, hence, $hn(y)$ is not in S. $\qquad\square$

We are now ready for the main result.

Theorem 4.21.   *A set is creative iff it is 1-complete iff it is m-complete.*

Proof.   By Theorem 4.20, every creative set is 1-complete. By definition, every 1-complete set is m-complete. By Theorem 4.16, every m-complete set is creative. $\qquad\square$

We conclude this section by mentioning a few additional facts without any detailed discussion.

A recursively enumerable set is **simple** iff its complement is infinite but has no infinite recursively enumerable subset. An example of a simple set is given in Exercise 4.9. It is straightforward to show that a simple set is never creative. (Hence, recursive and creative sets do not cover all recursively enumerable sets.) It is also clear that the complement of a simple set is neither recursively enumerable nor productive. (Hence, the latter two types of sets do not cover all sets, although there are nondenumerably many productive sets.)

As pointed out earlier, the relations $\leq_m$ and $\leq_1$ are referred to as *strong reducibilities*. Although very natural from the mathematical point of view, these relations do not in all cases reflect the intuitive idea of reducibility in a satisfactory fashion. A typical example is that the sets SA and $\sim SA$ are incomparable although, from the intuitive point of view, an algorithm for solving the membership in SA should yield an algorithm for solving the membership in $\sim SA$, and vice versa. This disadvantage is removed if *truth-table reducibility* is considered.

A set S is **truth-table reducible** to a set S'—in symbols, $S \leq_{tt} S'$—iff for any integer $x \geq 0$ we can find effectively finitely many integers $y_1, \ldots, y_t$ (here t may be different for different x) and a method by which the question $x \in S$? can be correctly answered if the correct answers to the questions $y_1 \in S'$?, $\ldots$, $y_t \in S'$? are known. The formal details of this definition, based on the idea of encoding the t questions as one integer, are omitted.

A further modification of the notion of reducibility leads to *Turing reducibility*. (An example of a case where truth-table reducibility is not intuitively sufficient is given in Exercise 4.10.) A Turing machine *TM* is provided with an **oracle for a set** S'. This means that during its computation, *TM* may ask questions about whether or not the number on its tape at a specific moment during the computation actually is in S'. This is formalized by letting *TM* go to the state q' (resp. q'') when scanning the letter a in the state q, depending on whether the number on its tape is (resp. is not) in S'. If such a Turing machine *TM* with an oracle for the set S' computes the characteristic function for the set S, we say that S is **Turing reducible** to S'—in symbols, $S \leq_T S'$.

The reason why $\leq_T$ is more general than $\leq_{tt}$ is that questions concerning the membership in S' have to be asked *in advance* as far as $\leq_{tt}$ is concerned, whereas similar questions can be asked *during the computation* as far as $\leq_T$ is concerned. The reducibilities according to $\leq_{tt}$ and $\leq_T$ are referred to as **weak reducibilities.**

It is not difficult to prove that for all S and S',

$$S \leq_1 S' \quad \text{implies} \quad S \leq_m S' \quad \text{implies} \quad S \leq_{tt} S' \quad \text{implies} \quad S \leq_T S'.$$

Moreover, even for recursively enumerable sets, none of the implications is reversible.

Since reduction according to $\leq_T$ is very efficient, it is not easy to exhibit many different Turing degrees of recursively enumerable sets. (The notion of a degree is defined in the same way as before.) In fact, for a long time it was an open problem whether or not there are incomparable recursively enumerable Turing degrees. Such degrees can be exhibited; the interested reader is referred to [Ro].

Although Turing degrees may be incomparable, one can delineate an ordered subset of Turing degrees in a natural fashion and thus obtain some useful reference points. Let us denote the degrees in this subset by nonnegative integers. In regard to language theory, this ordering gives detailed information concerning undecidability.

In particular, degree 0 consists of decidable problems—that is, of recursive sets. Degree 1 consists of problems in the same Turing degree as SA and $HALT$. Examples are the emptiness of recursively enumerable languages and the equivalence of context-free languages. For degrees higher than 1, neither the set itself nor its complement is recursively enumerable. Examples of problems of degree 2 are the infinity and equivalence problem for recursively enumerable languages. Another example is the problem of determining whether or not a context-free language is regular. Examples of problems of degree 3 are the problem of determining whether an arbitrary grammar G generates a regular (or a context-free) language and whether the complement of $L(G)$ is recursively enumerable.

4.6. UNIVERSALITY IN TERMS OF COMPOSITION

We have seen in Theorem 4.3 that all partial recursive functions of m variables can be expressed in terms of one particular partial recursive function $f_u(i, y_1, \ldots, y_m)$ by letting the first variable i range through all indices. We shall now establish a different universality result: There is an effectively obtainable partial recursive function $f(x, y)$ of two variables such that every partial recursive function (of arbitrarily many variables) can be expressed as a composition of f. We allow arbitrarily "deep" compositions, where the variables can be used in an arbitrary fashion. On the other hand, constants cannot be used as such in compositions.

More specifically, consider an arbitrary partial function $f(x, y)$ (not necessarily partial recursive). A **composition sequence** of f is any word consisting of f, variables, parentheses, and commas, which can be formed by finitely many applications of the Rules 1–2.

Rule 1. Each variable alone is a composition sequence of f.
Rule 2. If α and β are composition sequences of f, then so is $f(\alpha, \beta)$.

For instance,

$$f(f(x, x), x), \qquad f(x, f(y, z)), \qquad f(f(f(x, y), x), f(y, x))$$

are all composition sequences of F. We assume that all the variables occurring in composition sequences come from some fixed denumerable collection of variables.

A partial function g (of arbitrarily many variables) is **generated** by a partial function f iff g equals a composition sequence of f. For instance, if

$$g(x, y, z) = f(x, f(y, z)),$$

then g is generated by f. As before, equations for partial functions mean that both sides are defined for the same arguments and, whenever defined, assume the same value. A composition sequence σ is defined for a certain argument iff each partial sequence $f(\alpha, \beta)$ appearing in σ is defined for this argument or the appropriate part of it.

We assume that the variables of all partial functions considered in this section range over some subset of the set N of positive integers and that the function values are contained in N. This assumption is not essential and is made only because we want to use the pairing function defined in Section 4.3. If we want to include the number 0, for example, in the domains and ranges, the pairing function has to be modified accordingly.

Consider, thus, the pairing function φ and its generalizations φ_i, as defined in Section 4.3. For every partial function g of i variables, there is a partial function g_1 of one variable (obtainable effectively from g and referred to as the **one-place function corresponding to g**) such that

$$g(x_1, \ldots, x_i) = g_1(\varphi_i(x_1, \ldots, x_i)).$$

Clearly, if a partial function f generates some functions, it also generates an arbitrary composition of them. Since all the generalizations φ_i were defined as compositions of φ, we obtain the following result.

Lemma 4.22. *If a partial function generates the pairing function φ and the one-place function g_1 corresponding to a partial function g, it generates g itself.*

The next lemma is our basic tool in the main construction.

Lemma 4.23. *Any two partial functions $g(x, y)$ and $h(x, y)$ are generated by a single partial function $f(x, y)$.*

Proof. Let g_1 and h_1 be the one-place functions corresponding to g and h. By Lemma 4.22, it suffices to construct a partial function $f(x, y)$ generating the three functions g_1, h_1, and φ. Such a function $f(x, y)$ can be defined as follows. (The notation α^i stands for i applications of α for all i.)

$$f(x, x) = x + 1 \quad \text{for} \quad 1 \leq x \leq 4, \qquad f(x, x) = 1 \quad \text{for} \quad x \geq 5;$$
$$f(x, \alpha(x)) = g_1(x) \qquad \text{where} \quad \alpha(x) = f(x, x),$$
$$f(x, \alpha^2(x)) = h_1(x),$$
$$f(x, \alpha^3(x)) = 2x + 5, \qquad f(x, \alpha^4(x)) = 2x + 4;$$
$$f(2x + 5, 2y + 4) = \varphi(x, y);$$
$$f(x, y) = 1 \qquad \text{in all other cases.}$$

We first show that no contradiction arises in this definition—that is, that there is no argument (x, y) for which the value $f(x, y)$ has been defined in two different ways. For every x, the numbers x, $\alpha(x)$, $\alpha^2(x)$, $\alpha^3(x)$, $\alpha^4(x)$ are all distinct. This implies that all the arguments of f appearing on the first four lines of the definition are distinct. Since $\alpha(x) \leq 5$ for all x, we conclude that every pair $(2x + 5, 2y + 4)$ is distinct from all pairs of the form $(z, \alpha^i(z))$, for $i = 1, 2, 3, 4$. Finally, because $2x + 5$ is always odd and $2y + 4$ is always even, we see that every pair $(2x + 5, 2y + 4)$ is also distinct from all pairs (z, z).

The definition of $f(x, y)$ is given in such a way that we can immediately see that f generates the functions $\alpha(x)$, $g_1(x)$, $h_1(x)$, $2x + 5$, $2x + 4$, and $\varphi(x, y)$.   □

If g and h are partial recursive, then f is also partial recursive and effectively obtainable from g and h. Indeed, the only values for which $f(x, y)$ might be undefined are the values $f(x, \alpha(x))$ and $f(x, \alpha^2(x))$. These values are computed from the values of $g_1(x)$ and $h_1(x)$.

Observe also that there is, in fact, a continuum of functions $f(x, y)$ satisfying Lemma 4.23 for given g and h. This follows because the last line of the definition of f in the proof is arbitrary in the sense that the values are not needed when generating g and h.

It is clear that no nondenumerable collection of functions can be generated by a single function f because, clearly, all composition sequences of f can be numerated. On the other hand, the next theorem shows that every denumerable collection of functions can indeed be generated by a single function.

Theorem 4.24.   *For every finite or denumerable set K of partial functions (the number of variables may be different for different functions), there is a partial function $f(x, y)$ generating every function in K.*

Proof.   Let $k_i(x)$, $i = 1, 2, \ldots$, be the one-place functions corresponding to the functions in K. Let $h(x, y)$ be any partial function such that

$$h(x, x) = x + 1,$$

$$h(x, x + i) = k_i(x).$$

for all positive integers x and i. Clearly, h generates all partial functions $k_i(x)$.

We now choose in Lemma 4.23 $g = \varphi$ and h as above. Theorem 4.24 then follows by Lemmas 4.22 and 4.23.   □

Of course, in general the function $f(x, y)$ in Theorem 4.24 cannot be found effectively. However our final theorem is now an immediate consequence of Theorem 4.24, by the remarks made after Lemma 4.23. Indeed, it is obvious that all partial recursive functions form a denumerable set K for which the values of the corresponding one-place functions $k_i(x)$ can be found effectively by dovetailing.

Theorem 4.25. *There is an effectively constructable partial recursive function $f(x, y)$ generating every partial recursive function (of arbitrarily many variables).*

EXERCISES

4.1.   Compute the first few values $g(1)$, $g(2)$, $g(3)$, ... of the function $g(k)$ discussed in Example 4.1. What is the smallest value of k for which you find the computation intractable? (You may try to guess the correct Turing machine for a given k but should be aware of the fact that amazingly complicated Turing machines exist even for $k = 4$.)

4.2.   Explain why the given proof of Theorem 4.4 is better than the simpler arguments indicated after the proof, if one wants to eliminate the usage of Church's thesis. Try to prove Theorem 4.4 by applying unary functions in the definition of τ.

4.3.   Establish the undecidability of Problems (i)–(vi) mentioned after Theorem 4.6 without using Rice's theorem.

4.4.   Establish Theorem 4.7 without using the recursion theorem.

4.5.   Show that, for some i, $S_i = \{i, i^2\}$.

4.6.   The pairing functions φ_n introduced in Section 4.3 enable us to extend statements concerning sets into statements concerning relations. An n-ary relation $R(x_1, \ldots, x_n)$ can be encoded as the set $\{\varphi(x_1, \ldots, x_n) \mid R(x_1, \ldots, x_n)$ holds$\}$. If Q is a property defined for sets, we say that R has the property Q iff the encoding of R has the property Q. Thus, we may speak of recursive and recursively enumerable relations. Prove that an arbitrary function $f(x)$ is recursive iff f viewed as a binary relation is recursive iff f viewed as a binary relation is recursively enumerable.

4.7.   A relation $R_1(x_1, \ldots, x_{n-1})$ is termed the *projection* of a relation $R(x_1, \ldots, x_n)$ along the nth coordinate iff $R_1(x_1, \ldots, x_{n-1})$ holds exactly in case $R(x_1, \ldots, x_{n-1}, x_n)$ holds for some x_n. Projections along the other coordinates are defined analogously. Prove that a relation is recursively enumerable iff it is a projection of some recursive relation.

4.8.   Give a detailed proof for Theorem 4.16.

4.9.   Prove that the set S defined as follows is simple. First consider the set

$$S' = \{\varphi(x, y) \mid y \in S_x \text{ and } y > 2x\}.$$

It is easy to see that the set S' is recursively enumerable. Consider some fixed effective enumeration of S' and define the set S'' as follows. A number $t = \varphi(x, y)$ is in S'' iff (1) $t = \varphi(x, y)$ is in S' and (2) whenever $\varphi(x, z)$ is in S' with $z \neq y$, then $\varphi(x, z)$ comes after $\varphi(x, y)$ in the enumeration. Finally, S equals the set of all numbers y such that for some x, $\varphi(x, y)$ is in S''.

4.10. Intuitively, a set S is reducible to a set S' iff any algorithm for solving the membership problem of S' yields an algorithm for solving the membership problem of S. Show by an example that truth-table reducibility does not satisfactorily represent the intuitive notion. ([Ro] contains more information.)

CHAPTER 5

Famous Decision Problems

5.1. POST CORRESPONDENCE PROBLEM AND APPLICATIONS

Many decision problems have already been discussed in previous chapters, and several undecidability results were established in Chapter 4. Chapter 5 deals with decision problems and undecidability results of a somewhat different nature. Indeed, we want to give a broader view of decision problems because undoubtedly decidability, as opposed to undecidability (we could also speak of solvability versus unsolvability or computability versus noncomputability) constitutes the chief issue in the theory of computation.

It was pointed out in Section 4.3 that proofs of undecidability are either direct or indirect. Indirect proofs require some points of reference: problems already known to be undecidable. In regard to language theory, the most useful among such points of reference is the Post correspondence problem. For instance, when dealing with context-free languages, it is rather difficult to use the halting problem as a point of reference, whereas the Post correspondence problem is very suitable.

The Post correspondence problem can be described as follows. We have to decide for two lists, both consisting of n words, whether or not some catenation of words in one list equals the **corresponding** catenation of words in the other list. Here *corresponding* means that words in exactly the same positions in the two lists must be used. For example, if the two lists are

$$(a^2, b^2, ab^2) \quad \text{and} \quad (a^2b, ba, b) \tag{1}$$

and we catenate the first, second, first, and third words in the lists (in this order), then the word $a^2b^2a^3b^2$ results in both cases. We express this fact by saying that the sequence of indices 1, 2, 1, 3, is a *solution* for instance (1) of the Post correspondence problem. On the other hand, the instance consisting of the two lists

$$(a^2b, a) \quad \text{and} \quad (a^2, ba^2) \tag{2}$$

possesses no solution. This can be shown by the following argument. An eventual solution of (2) must begin with the index 1. Then the index 2 must follow because the word coming from the second list must "catch up" the missing b. So far, we have the words a^2ba and a^2ba^2. Since one of the words must always be a prefix of the other, we conclude that the next index in the eventual solution is 2, yielding the words a^2ba^2 and $a^2ba^2ba^2$. But because neither of the words in the first list begins with b, we conclude that no futher continuation is possible.

The notion of a morphism provides a convenient way of giving a formal definition for the Post correspondence problem.

Definition 5.1. *Let g and h be two morphisms with the domain alphabet Σ and range alphabet Δ. The pair (g, h) is called an* **instance of the Post correspondence problem.** *A nonempty word w over Σ such that $g(w) = h(w)$ is referred to as a* **solution** *of this instance.* $\square$

Clearly, the two lists mentioned in the discussion preceding our definition are obtained by listing the values $g(a)$ and $h(a)$, where a ranges over Σ.

In some instances of the Post correspondence problem, it is rather easy to find out whether or not they have solutions. We observed above that (1) has a solution, whereas (2) has no solution. There are also some classes of instances for which an easy decision procedure exists. Such a class consists of all instances for which Δ consists of a single letter. Then the two lists have the form

$$(a^{i_1}, \ldots, a^{i_n}) \quad \text{and} \quad (a^{j_1}, \ldots, a^{j_n}) \tag{3}$$

It is easy to verify that (3) has a solution iff either there is a t such that $i_t = j_t$, or else there are t and u such that $i_t > j_t$ and $i_u < j_u$.

By the decidability of the Post correspondence problem, we mean the existence of an algorithm that settles *every* instance; that is, given an arbitrary instance as its input, the algorithm produces an answer of yes or no, depending on whether or not the instance has a solution. Although some classes of instances possess a decision method, there can be no decision method for the general problem.

Theorem 5.1. *The Post correspondence problem is undecidable.*

Proof. By Theorem 4.8, recursively enumerable languages (as defined in Chapter 2) coincide with languages acceptable by Turing machines. Hence, if we can solve the membership problem for recursively enumerable languages, we can solve the halting problem, which is impossible by Theorem 4.4. To prove Theorem 5.1, we show that an algorithm for solving the Post correspondence problem yields an algorithm for solving the membership problem of recursively enumerable languages.

Assume that the Post correspondence problem is decidable and consider an arbitrary gammar G with the nonterminal and terminal alphabet Σ_N and Σ_T, the start letter S, and the set P of productions. Further, consider an arbitrary word w over Σ_T. We construct an instance, PCP, of the Post correspondence problem such that PCP has a solution iff w is in $L(G)$. Since the construction of PCP will be effective, this gives a method of deciding whether or not w is in $L(G)$.

By Theorem 2.3, we may assume that G has no production whose left side equals λ. (We need only this part of Theorem 2.3—we do not need the normal-form result in its full strength.) We may also assume that G has no production whose right side equals λ. For if this is not the case originally, we then replace every production $\alpha \to \lambda$ by all productions $a\alpha \to a$ and $\alpha a \to a$, where a ranges over the letters of $\Sigma_N \cup \Sigma_T$. In this process we may lose the empty word from the language generated, but otherwise the new grammar generates the same language as the old one. On the other hand, this eventual loss of the empty word does not affect the decidability results underlying our argument.

We denote the letters of $\Sigma_N \cup \Sigma_T = \Sigma_1$ by $a_1, \ldots, a_r$, where a_1 is the start letter. (To obtain a uniform notation for the letters, we abandon the convention in which small letters are terminals.) Let

$$\{\alpha_i \to \beta_i \mid 1 \le i \le n\}$$

be the set of productions of G. We know that none of the words α_i and β_i are empty. We let

$$\Sigma_1' = \{a' \mid a \in \Sigma_1\} \quad \text{and} \quad \Delta = \Sigma_1 \cup \Sigma_1' \cup \{B, E, \#, \#'\}.$$

(It is assumed that B, E, $\#$ and $\#'$ are letters not contained in the other alphabets considered. Intuitively, B marks the beginning of a derivation and E the end of it, whereas $\#$ and $\#'$ are markers between two consecutive words in a derivation.) For any word x over Σ_1, we denote by x' the word over Σ_1' obtained from x by replacing every letter a_i with a_i'.

Consider the instance $PCP = (g, h)$ of the Post correspondence problem. The domain alphabet of the morphisms g and h is

$$\Sigma = \{1, 2, \ldots, 2r + 2n + 4\},$$

and the range alphabet is the alphabet Δ defined above. The morphisms themselves are defined by Table 5.1, where i ranges over the numbers $1, \ldots, r$ and j ranges over the numbers $1, \ldots, n$.

TABLE 5.1

	1	2	3	4	$4+i$	$4+r+i$	$4+2r+j$	$4+2r+n+j$
g	$Ba_1\#$	$\#'$	$\#$	E	a_i'	a_i	β_j'	β_j
h	B	$\#$	$\#'$	$\#'wE$	a_i	a_i'	α_j	α_j'

We claim that *PCP* possesses a solution iff $w \in L(G)$. Hence, if the Post correspondence problem is decidable, then so is the membership problem for recursively enumerable languages.

The idea behind the construction of *PCP* is that derivations according to G will be simulated. Moreover, g produces the derivation faster than h because it starts with $Ba_1\#$, whereas h starts only with B. The morphism h can catch up at the end of the derivation (by the index 4) iff w is the last word of the derivation.

For example, assume that the first three productions of G are (in this order)

$$S \to aCD, \qquad CD \to D, \qquad D \to d.$$

Assume, further, that a and D are the second and third letters of Σ_1, respectively. (Recall that $S = a_1$.) Consider the word $w = ad$. Clearly, w has the following derivation:

$$S \Rightarrow aCD \Rightarrow aD \Rightarrow ad. \tag{4}$$

Then a solution for *PCP* is constructed according to Table 5.2. Hence, the word

$$1(4 + 2r + 1)2(4 + r + 2)(4 + 2r + n + 2)3(4 + 2)(4 + 3)2(4 + r + 2)(4 + 2r + n + 3)4$$

(where parentheses have been added for clarity) is a solution for *PCP*, the common value of the morphisms g and h being for this word:

$$BS\#a'C'D'\#'aD\#a'D'\#'adE. \tag{5}$$

TABLE 5.2

	1	$4+2r+1$	2	$4+r+2$	$4+2r+n+2$	3	$4+2$
g	$BS\#$	$a'C'D'$	$\#'$	a	D	$\#$	a'
h	B	S	$\#$	a'	$C'D'$	$\#'$	a

	$4+3$	2	$4+r+2$	$4+2r+n+3$	4
g	D'	$\#'$	a	d	E
h	D	$\#$	a'	D'	$\#'adE$

Clearly, (5) results from (4) after the following simple modifications. The boundary markers B and E have been added to the beginning and end. The sign $\Rightarrow$ has been replaced by $\#$ and, moreover, every second step has been primed, as has every second marker $\#$. Finally, an "idle" step (from aD to $a'D'$) has been added in order to reach the nonprimed word ad corresponding to the final index of (4).

The reader should have no difficulty in verifying that an analogous argument is also valid in general: Whenever w is in $L(G)$, then PCP has a solution. The interested reader may consult [Sa3], where this fact is established by a detailed inductive argument.

It is somewhat more difficult to establish the converse—that is, to show that if PCP has a solution, then w is in $L(G)$. Observe first that primed versions are introduced for every second step in a derivation in order to avoid unwanted solutions: Without primes the indices $4 + i$ would yield a solution immediately. By checking through all indices, it is immediately verified that every solution must begin with the index 1 and end with the index 4. (Recall that the words α_i and β_i are nonempty.) This gives a suitable starting point for an inductive argument establishing the converse. The details are left to the reader. Also, [Sa3] can be consulted.   □

Theorem 5.1 can be strengthened in the following way. Assume that we consider only such instances of the Post correspondence problem where the range alphabet of the morphisms equals $\{a, b\}$. We claim that there is no algorithm for deciding whether or not such an instance of the Post correspondence problem possesses a solution. Assume the contrary. Consider then an arbitrary instance PCP of the Post correspondence problem, where the morphisms g and h have the range alphabet $\Delta = \{a_1, \ldots, a_r\}$. Encode the letters of Δ by the morphism f:

$$f(a_i) = ba^i b \qquad \text{for } i = 1, \ldots, r,$$

and consider the instance $PCP' = (fg, fh)$ of the Post correspondence problem. Clearly, the range alphabet of PCP' is $\{a, b\}$. It is also easy to verify that PCP' possesses a solution iff PCP possesses a solution. Hence, by our assumption, the Post correspondence problem would be decidable, which contradicts Theorem 5.1. Therefore, we have established the following result.

Theorem 5.2.   *There is no algorithm for solving instances (g, h) of the Post correspondence problem in which the range alphabet of the morphisms is $\Delta = \{a, b\}$.*

We have seen that the Post correspondence problem is decidable for all such instances where the range alphabet consists of one letter only. Thus, the decidability status is understood as far as the range alphabet is concerned.

The situation is entirely different with respect to the domain alphabet. Trivially, the case in which the domain alphabet consists of one letter only is decidable. For a long time, the case in which the domain alphabet consists of two letters was a celebrated open problem, until decidability was finally established in [EhKR]. Otherwise, little is known about the decidability status as far as the domain alphabet is concerned.

Let us define the **bad-luck number** to be the smallest integer k such that the Post correspondence problem is undecidable even if only such instances in which the cardinality of the domain alphabet of the two morphisms is less than or equal k are considered.

To show that the bad-luck number exists (and, thus, is a specific unique integer), we first observe that there is a specific (effectively constructable) Turing machine TM_{i_0} for which it is undecidable whether or not it halts with a given input x. This is a consequence of the following stronger version of Theorem 4.4.

Lemma 5.3. *There exists (effectively) an index i_0 for which the function*

$$g(x) = \begin{cases} 1 & \text{if } TM_{i_0}(x){\downarrow} \\ 0 & \text{if } TM_{i_0}(x){\uparrow} \end{cases}$$

is not recursive.

Proof. Let i_0 be an index for the following Turing machine. The machine first converts an arbitrary input z into the pair $(\psi_1(z), \psi_2(z))$ (where ψ_1 and ψ_2 are the inverses of the pairing function discussed in Section 4.3) and then calls the universal machine for functions of one variable, described in Theorem 4.3. If our function g would be recursive for this i_0, we would be solving the general halting problem, which would contradict Theorem 4.4. □

Consider the proof of Theorem 5.1. By Lemma 5.3, we may restrict our attention only to such instances of the Post correspondence problem where the morphisms correspond to the Turing machine TM_{i_0} and, consequently, have a fixed domain alphabet. The bad-luck number must be less than or equal to the cardinality of this alphabet.

The estimate obtained in this fashion for the bad-luck number is very rough. The best known estimate, in [Cl] and [Pa], shows that the bad-luck number is less than or equal 9. Hence, we know that it is between 3 and 9.

As already pointed out, the Post correspondence problem is a very useful tool, especially for establishing language-theoretic undecidability results. This is illustrated by our next three theorems. Further material is contained in Exercise 5.2.

Theorem 5.4. *There is no algorithm for deciding whether or not two given context-free languages intersect.*

Proof. Assuming the existence of such an algorithm, we show the decidability of the Post correspondence problem, which contradicts Theorem 5.1.

Consider an arbitrary instance $PCP = (g, h)$ of the Post correspondence problem, where the morphisms have the domain alphabet $\Sigma = \{1, \ldots, n\}$ and are defined by

$$g(i) = \alpha_i \quad \text{and} \quad h(i) = \beta_i \quad \text{for } i = 1, \ldots, n.$$

(Here α_i and β_i are words over some alphabet Δ disjoint with Σ.)

Consider the context-free grammar G_α (resp. G_β), determined by the productions

$$S \to iS\alpha_i \quad \text{and} \quad S \to i\alpha_i \quad (\text{resp. } S \to iS\beta_i \text{ and } S \to i\beta_i),$$

where $i = 1, \ldots, n$ and S is the only nonterminal. Clearly, PCP possesses a solution iff the intersection of $L(G_\alpha)$ and $L(G_\beta)$ is nonempty. $\square$

In the previous proof, the application of Theorem 5.1 is quite straightforward. A more-involved reduction is applied in our next proof.

Theorem 5.5.   *The equivalence problem for context-free grammars is undecidable.*

Proof.   Consider again an arbitrary instance $PCP = (g, h)$ of the Post correspondence problem. The notations Σ, Δ, α_i, and β_i are defined exactly as in the previous proof.

Three languages L, L_g, and L_h over the alphabet $\Sigma \cup \Delta \cup \{\#\}$, where $\#$ is a new letter, are now introduced. By definition,

$$L = \Sigma^* \# \Delta^*.$$

The language L_g is the subset of L consisting of all words in L, with the exception of words of the form

$$w \# g(\text{mi}(w)), \quad w \neq \lambda. \tag{6}$$

(As before, mi denotes the mirror image.) The language L_h is defined in the same way, with h instead of g in (6).

Clearly,

$$PCP \text{ has no solution} \quad \text{iff} \quad L = L_g \cup L_h. \tag{7}$$

We now define two context-free grammars G_1 and G_2 as follows. Both grammars have the start letter S and the nonterminal and terminal alphabets

$$\{S, S_1, S_2, S_3, S_4, A, B\} \quad \text{and} \quad \Sigma \cup \Delta \cup \{\#\}.$$

The productions of G_1 are listed below. In the list it is understood that a runs through Σ and b through Δ.

$$S \rightarrow A, S \rightarrow B, \quad S \rightarrow S_4, \quad S \rightarrow \#,$$
$$A \rightarrow aAg(a), \quad B \rightarrow aBh(a),$$
$$A \rightarrow S_3 b, \quad B \rightarrow S_3 b,$$
$$A \rightarrow aS_1 x, \quad B \rightarrow aS_1 y,$$

where x (resp. y) runs through all words over Δ shorter than $g(a)$ (resp. $h(a)$), including the empty word;

$$A \rightarrow aS_2 x, B \rightarrow aS_2 y,$$

where x (resp. y) runs through all words over Δ of the length $|g(a)|$ (resp. $|h(a)|$) but different from $g(a)$ (resp. $h(a)$);

$$S_1 \rightarrow aS_1, \quad S_1 \rightarrow \#,$$
$$S_2 \rightarrow aS_2, \quad S_2 \rightarrow S_3,$$
$$S_3 \rightarrow S_3 b, \quad S_3 \rightarrow \#,$$
$$S_4 \rightarrow aS_4, \quad S_4 \rightarrow S_4 b.$$

The productions of G_2 are obtained from those of G_1 by replacing the two productions $S_1 \rightarrow \#$ and $S_3 \rightarrow \#$ with the production $S_4 \rightarrow \#$. This concludes the definition of G_1 and G_2.

We claim that

$$L(G_1) = L_g \cup L_h \quad \text{and} \quad L(G_2) = L. \tag{8}$$

In fact, the second equation is obvious: L is generated according to G_2 via the nonterminals S and S_4 because no terminal word results if A or B is introduced.

The proof of the first of Equations (8) is more involved. The language L_g is generated via the nonterminal A and L_h via B. As long as the derivation stays in A, there is a correct match in the sense of (6). The introduction of S_1 indicates that the part of the word coming after the center marker # is too short. Similarly, the introduction of S_3 directly from A or B indicates that the part is too long. The introduction of S_2 indicates that an unrepairable error has been created in the matching. By these remarks, the reader should be able to establish the first equation in (8). Also, [Sa4] may be consulted.

By (7) and (8), an algorithm for solving the equivalence problem for context-free grammars yields an algorithm for solving the Post correspondence problem. □

In the preceding proof, the grammars G_1 and G_2 contain some superfluous productions because some of the nonterminals never lead to a terminal word. The reason for this is that we also want to use the same proof to establish our next theorem.

Every word derivable from the start letter of a grammar G is referred to as a **sentential form** of G. Thus, sentential forms may contain both nonterminals and terminals.

Observe now that the grammars G_1 and G_2 in the preceding proof generate exactly the same sentential forms iff $L(G_1) = L(G_2)$. This follows because the sentential forms containing nonterminals coincide in both grammars.

Observe also that if the production $c \to c$ is added for every terminal letter c, then the resulting grammars, viewed as 0L systems, are equivalent iff the original grammars G_1 and G_2 generate the same language. Hence, we have established the following result.

Theorem 5.6. *The equivalence problem for 0L systems is undecidable. There is no algorithm for deciding whether or not two given context-free grammars generate the same sentential forms.*

5.2. HILBERT'S TENTH PROBLEM AND CONSEQUENCES: MOST QUESTIONS CAN BE EXPRESSED IN TERMS OF POLYNOMIALS

The topic of this section can justly be referred to as the most famous specific decision problem of all time. *Hilbert's tenth problem* consists of giving an algorithm that will tell whether or not a given polynomial equation with integer coefficients has a solution in integers.

As it turns out, no such algorithm exists—that is, the problem is undecidable. It seems possible, although rather surprising, that Hilbert already anticipated this state of affairs when he stated the problem as the tenth in his list of problems proposed at the International Congress of Mathematicians in 1900. When commenting on the proof of the impossibility of the solution of mathematical problems, Hilbert stated [Hi]: "Sometimes it happens that we seek the solution under unsatisfied hypotheses or in an inappropriate sense and are therefore unable to reach our goal. Then the task arises of proving the impossibility of solving the problem under the given hypotheses and in the sense required. Such impossibility proofs were already given by the ancients, e.g., that the hypotenuse of an isosceles right triangle has an irrational ratio to its leg. In modern mathematics the question of the impossibility of certain solutions has played a key role, so that we have acquired the knowledge that such old and difficult problems as to prove the parallel axiom, to square the circle, or to solve equations of the fifth degree in radicals have no solution in the originally intended sense, but nevertheless have been solved in a precise and completely satisfactory way." Such a "precise and completely satisfactory" solution to Hilbert's tenth problem was provided by Matijasevič in 1970: undecidable. Of course, nothing was known about undecidability in 1900 (or even much later) and, in view of Hilbert's attempts to find general decision methods in logic, it remains questionable whether the quoted statement was also intended to concern the tenth problem.

Thus, we consider equations of the form

$$P(x_1, \ldots, x_n) = 0, \tag{9}$$

where P is a polynomial in the variables $x_1, \ldots, x_n$ with integer coefficients. Hilbert's tenth problem consists of deciding whether or not an arbitrary equation (9) has a solution in integers. However, from the decidability point of view, it is equivalent to ask whether or not (9) has a solution in nonnegative integers. This can be seen by the following argument. Because every nonnegative integer is a sum of four squares, equation (9) has a solution in nonnegative integers iff the equation

$$P(p_1^2 + q_1^2 + r_1^2 + s_1^2, \ldots, p_n^2 + q_n^2 + r_n^2 + s_n^2) = 0$$

has an integral solution. On the other hand, (9) has a solution in integers iff the equation

$$P(p_1 - q_1, \ldots, p_n - q_n) = 0$$

has a solution in nonnegative integers.

In a similar way, it is seen that—from the decidability point of view—it is also equivalent to ask whether or not (9) has a solution in positive integers. Of course, the set S of solutions is recursively enumerable and, consequently, there is an effective procedure for listing S. To decide whether or not S is empty, we would have to know when to stop the listing procedure if so far no elements of S have been produced. In some cases, however, we would have to wait for quite a long time. For instance, the smallest positive solution of the equation $x^2 = 991y^2 + 1$ is

$$x = 379516400906811930638014896080,$$

$$y = 12055735790331359447442538767.$$

We now begin the formal details, first introducing the most important notion.

Definition 5.2. *A set S of ordered n-tuples, $n \geq 1$, of nonnegative integers is termed **Diophantine** iff there is a polynomial $P(x_1, \ldots, x_n, y_1, \ldots, y_m)$ with integer coefficients (which may be negative) such that for all nonnegative integers $x_1, \ldots, x_n$,*

$$(x_1, \ldots, x_n) \in S \quad \text{iff} \quad (\exists y_1 \geq 0) \ldots (\exists y_m \geq 0)[P(x_1, \ldots, x_n, y_1, \ldots, y_m) = 0]\square$$

For $n > 1$ we usually speak of **Diophantine relations** instead of sets of n-tuples. The notion of recursive enumerability is extended in the natural way to concern relations. An n-ary **relation** R is **recursively enumerable** iff the set

$$\{\varphi_n(x_1, \ldots, x_n) \mid R(x_1, \ldots, x_n) \text{ holds}\},$$

where φ_n is the extension of the pairing function, is recursively enumerable. (Whenever we are dealing with nonnegative rather than positive integers, the pairing function is modified accordingly.)

Some simple examples of Diophantine sets and relations follow. As in the definition, the variables associated to existential quantifiers range over nonnegative integers.

The set S_{comp} of composite numbers is Diophantine because

$$x \in S_{comp} \quad \text{iff} \quad (\exists y, z)[x = (y + 2)(z + 2)].$$

The ordering relations, as well as the divisibility relation, are Diophantine because

$$x < y \quad \text{iff} \quad (\exists z)[x + z + 1 = y],$$
$$x \leq y \quad \text{iff} \quad (\exists z)[x + z = y],$$
$$x \mid y \quad \text{iff} \quad (\exists z)[xz = y].$$

There are various straightforward techniques for showing that certain relations are Diophantine. For instance, the language of Diophantine relations permits the use of existential quantifiers (this is clear by the definition), as well as the logical connectives *and* and *or*. If we want to say that both $P_1 = 0$ and $P_2 = 0$ (resp. either $P_1 = 0$ or $P_2 = 0$), we may express this as $P_1^2 + P_2^2 = 0$ (resp. $P_1 P_2 = 0$). An extension to several polynomials instead of two is obvious.

On the other hand, the language of Diophantine relations permits neither the use of universal quantifiers nor the use of the logical negation. It turns out that every set that can possibly be Diophantine actually is Diophantine: All recursively enumerable sets are Diophantine. In view of this fact it is obvious that a universal quantification or negation may convert a Diophantine relation into a non-Diophantine one.

Diophantine sets are defined as sets for which a certain polynomial equation possesses a solution. It is often intuitively easier to deal with sets of values of a polynomial. Therefore, the following result is pleasing.

Theorem 5.7. *For every Diophantine set S of positive integers, there is a polynomial Q such that S equals the set of positive values of Q when the variables of Q range over nonnegative integers.*

Proof. By the definition of Diophantine sets, there is a polynomial P such that

$$x \in S \quad \text{iff} \quad (\exists y_1, \ldots, y_m)[P(x, y_1, \ldots, y_m) = 0]. \tag{10}$$

Define the polynomial Q by

$$Q(x, y_1, \ldots, y_m) = x[1 - (P(x, y_1, \ldots, y_m))^2].$$

By (10), every element of S is in the range of Q. Conversely, a positive integer x can be in the range of Q only if there are nonnegative integers $y_1, \ldots, y_m$ with the property $P(x, y_1, \ldots, y_m) = 0$, which shows that x is in S. $\qquad \square$

By definition, every Diophantine set S of nonnegative integers satisfies a condition of the form (10). Let us refer to the least degree of P for which (10) is satisfied as the **degree** of S and to the least value of m as the **dimension** of S. In general, the degree and dimension are not reached by the same polynomial because if one tries to reduce the degree of a polynomial, then one usually has to increase the number of variables, and vice versa. This is illustrated by the proof of the following theorem.

Theorem 5.8. *Every Diophantine set is of degree less than or equal to 4.*

Proof. Assume that S satisfies (10) for some P. We introduce new variables z_j and replace P by simultaneous equations of the form $z_j = t_1 t_2$, where the t's are original variables of P. For instance, if P is the polynomial $x^4 y_1 - 3y_2^2$, then for all x, clearly,

$$(\exists y_1, y_2)[x^4 y_1 - 3y_2^2 = 0]$$
$$\text{iff} \quad (\exists y_1, y_2, z_1, z_2)[z_1 = x^2 \quad \text{and} \quad z_2 = z_1^2 \quad \text{and} \quad z_2 y_1 = 3y_2^2].$$

By this technique of introducing new variables, it is obvious that every equation $P = 0$ can be replaced by simultaneous equations $P_1 = 0, \ldots, P_n = 0$, where each P_i is of degree less than or equal to 2. Then the polynomial $Q = P_1^2 + \cdots + P_n^2$, clearly of degree less than or equal to 4, can be used as the defining polynomial for S. $\qquad \square$

It is much more difficult to prove that the dimension of Diophantine sets is also bounded—that is, that there is an integer n such that every Diophantine set is of dimension less than or equal to n. This surprising fact is, however, an immediate consequence of Theorem 5.9. As an application, consider the sets

$$T_k = \{x \mid (\exists y_1, \ldots, y_k)[x = (y_1 + 2) \cdots (y_k + 2)]\}.$$

Thus, T_2 is the set of composite numbers, whereas T_k consists of numbers that can be represented as a product of k nontrivial factors. It is surely unexpected that all sets T_k can be given by a Diophantine definition with no more than n parameters!

Theorem 5.9. *Every recursively enumerable set and relation is Diophantine.*

We omit the proof of Theorem 5.9. The interested reader will find detailed, self-contained proofs in [Da2] and [Ru1], for instance. The proof consists of three basic parts. One shows first that every recursively enumerable set or relation can be expressed by applying existential and bounded universal quantifiers to a polynomial equation. Essentially, this goes back to the work of Gödel and makes use of the representation of recursively enumerable sets as ranges of so-called primitive recursive functions. Secondly, one obtains a generalized Diophantine representation, where some variables may appear as exponents for an arbitrary recursively enumerable set or relation. As the third part of the proof, one then shows that the exponential relation $x = y^z$ is Diophantine. This third part stayed as an open problem for quite a long time. It was known that it suffices to show the existence of a Diophantine relation, where one component grows exponentially in terms of some other component. Matijasevitš showed that $x = a_{2y}$, where $a_0, a_1, a_2, \ldots$ constitute the Fibonacci sequence, is such a relation.

The following strong results are almost immediate corollaries of Theorem 5.9. As before, the quantified variables range over nonnegative integers.

Theorem 5.10.   *There is a polynomial P_u with integer coefficients such that for all i, the ith recursively enumerable set S_i satisfies, for every x, the condition*

$$x \in S_i \quad \text{iff} \quad (\exists y_1, \ldots, y_m)[P_u(i, x, y_1, \ldots, y_m) = 0].$$

Proof.   Because the relation $x \in S_i$ is recursively enumerable (because the set $HALT$ discussed in Example 4.6 is recursively enumerable), the theorem is an immediate consequence of Theorem 5.9. $\square$

The polynomial P_u is referred to as a **universal** polynomial: A Diophantine representation for every recursively enumerable set is obtained in terms of this single polynomial by letting the first variable range over positive integers. This P_u corresponds to the definition of Diophantine sets. Of course, one may construct a universal polynomial $P_u{}'$ also corresponding to the modification presented in Theorem 5.7; by fixing the value of one variable in $P_u{}'$, we get a representation of an arbitrary recursively enumerable set of positive integers as the set of positive values of $P_u{}'$.

The number of variables in the polynomial P_u constitutes an upper bound m for the dimension of every Diophantine set. The best possible value of m is not known. While a straightforward construction following the main line of the proof gives the value $m = 50$, it is possible to reduce the upper bound m to 9. However, then the degree of P_u will be enormous. The degree 4 can be obtained for the number of variables $m = 58$. There is also a universal polynomial with 26 variables and degree 24. The reader is referred to [Jo] for trade-offs between the degree and the number of variables.

Theorem 5.11. *There is a polynomial P with integer coefficients and a special variable i satisfying the following condition. No algorithm exists for deciding whether or not an arbitrary Diophantine equation $P_{i_0} = 0$ possesses a solution in nonnegative integers, where P_{i_0} results from P by substituting a positive integer i_0 for the special variable i. Hence, Hilbert's tenth problem is undecidable.*

Proof. The polynomial P_u of Theorem 5.10 can be chosen as P. An algorithm described in the statement of Theorem 5.11 is impossible because it would, in fact, also constitute an algorithm for the emptiness problem of recursively enumerable sets. The third sentence of Theorem 5.11 is an immediate consequence of the second sentence: The problem is undecidable even if only polynomials $P_{i_0} = 0$ are considered.   □

In spite of Theorem 5.11, large classes of polynomial Diophantine equations are effectively solvable. There is an algorithm for deciding the solvability of all Diophantine equations with degree less than or equal 2. By Theorem 5.8, no such algorithm exists for all Diophantine equations with degree less than or equal 4. The existence of an algorithm for settling the question in regard to Diophantine equations with degree less than or equal 3 is an open problem.

As regards the number of variables, it is obvious that Hilbert's tenth problem is decidable if only polynomials with one variable are considered. It is known that the problem is undecidable for polynomials with no more than 9 variables.

The results presented above have really surprising consequences even from a purely number-theoretic point of view.

Example 5.1. According to Theorems 5.9 and 5.7, the set of prime numbers can be expressed as the set of positive values of a polynomial Q (with integer coefficients, as always in this section). What would such a Q look like? An example due to [JoSWW], which uses exactly the letters of the English alphabet, is the following:

$$
\begin{aligned}
Q(a, \ldots, z) = (k + 2)\{1 &- [wz + h + j - q]^2 \\
&- [(gk + 2g + k + 1)(h + j) + h - z]^2 \\
&- [2n + p + q + z - e]^2 - [16(k + 1)^3(k + 2)(n + 1)^2 + 1 - f^2]^2 \\
&- [e^3(e + 2)(a + 1)^2 + 1 - o^2]^2 - [(a^2 - 1)y^2 + 1 - x^2]^2 \\
&- [16r^2y^4(a^2 - 1) + 1 - u^2]^2 \\
&- [((a + u^2(u^2 - a))^2 - 1)(n + 4dy)^2 + 1 - (x + cu)^2]^2 \\
&- [n + 1 + v - y]^2 \\
&- [(a^2 - 1)l^2 + 1 - m^2]^2 - [ai + k + 1 - l - i]^2 \\
&- [p + l(a - n - 1) + b(2an + 2a - n^2 - 2n - 2) - m]^2 \\
&- [q + y(a - p - 1) + s(2ap + 2a - p^2 - 2p - 2) - x]^2 \\
&- [z + pl(a - p) + t(2ap - p^2 - 1) - pm]^2\}.
\end{aligned}
$$

Thus, questions about prime numbers are equivalent to questions about positive values of the above polynomial Q! For instance, the celebrated open problem about the existence of infinitely many twin primes can be stated as follows: Are there infinitely many positive values α of Q such that also $\alpha - 2$ is a value of Q? (As before, the variables of Q range over nonnegative integers.) $\square$

Most of the famous open problems deal with recursively enumerable sets. Consequently, they can be expressed as problems concerning positive values of certain (effectively constructable) polynomials. This *principle of the reduction of mathematical problems to problems concerning positive values of polynomials* will now be illustrated further by considering Fermat's celebrated last problem.

Below, we give a Diophantine representation for the relation $p_1{}^k + p_2{}^k = p_3{}^k$. For simplicity, we restrict the attention to the positive values of the variables. Further details of the construction can be found in [Ru1].

$$p_1^k + p_2^k = p_3^k \quad \text{iff} \quad (\exists q_1, q_2, q_3)(q_1 = p_1^k \text{ and } q_2 = p_2^k \text{ and } q_3 = p_3^k \text{ and}$$
$$q_1 + q_2 = q_3)$$
$$\text{iff} \quad (\exists a_1, a_2, a_3, b_1, b_2, b_3, \ldots, \tau_1, \tau_2, \tau_3)$$
$$\{\sum_{i=1}^{3} [(a_i - q_i - p_i - k - p_i^2 - 2)^2 + (u_i^2 - a_i(a_i + 2)u_i w_i + a_i^2 w_i^2 - 1)^2$$
$$+ (l_i - u_i - b_i)^2 + (l_i - v_i - c_i)^2 + (l_i^2 - l_i z_i - z_i^2 - 1)^2$$
$$+ (g_i^2 - g_i h_i - h_i^2 - 1)^2 + (g_i - d_i l_i^2)^2 + (m_i - 2 - e_i l_i)^2$$
$$+ (m_i - 3 - j_i(2h_i + g_i))^2 + (x_i^2 - m_i x_i y_i + y_i^2 - 1)^2$$
$$+ (x_i - u_i - n_i l_i)^2 + (x_i - v_i - r_i(2h_i + g_i))^2 + (s_i^2 - u_i s_i t_i + t_i^2 - 1)^2$$
$$+ (t_i - s_i - \theta_i)^2(s_i - t_i)^2 + (v_i - 9t_i - \sigma_i)^2 + (s_i - k - \rho_i(u_i - 2))^2$$
$$+ (t_i + (p_i - u_i)s_i - q_i - \tau_i(p_i u_i - p_i^2 - 1))^2] + (q_1 + q_2 - q_3)^2 = 0\}.$$

Similar polynomial representations can be given, for instance, for Riemann's and Goldbach's hypotheses. Gödel's famous incompleteness theorem can be strengthened to the following form. The result is on an intuitive level: We do not want to formalize the notion of an axiomatization.

Theorem 5.12. *For every axiomatization of the theory of natural numbers, there is a Diophantine equation with no solution in nonnegative integers such that the nonexistence of solutions cannot be proved within the axiomatization.*

Proof. Assume the contrary. Then Hilbert's tenth problem can be solved by running two enumeration procedures concurrently: (i) a procedure for listing all equations possessing solutions (this procedure consists of trial and error in a dovetailing fashion) and (ii) a procedure for listing all theorems to the effect that some Diophantine equation has no solution. $\square$

Consider equations $P = 0$, where P is a polynomial with integer coefficients, but the equation is not viewed as a Diophantine one—that is, noninteger values of the variables are also taken into account. According to Tarski's theorem, the elementary theory of real numbers is decidable. Consequently, there is an algorithm for deciding whether or not a given equation $P = 0$ has a solution in real numbers. The same problem concerning solutions in rational numbers is open. It is equivalent to the problem of finding an algorithm for deciding whether or not a given homogeneous equation $P = 0$ possesses a solution in integers.

5.3. WORD PROBLEMS AND VECTOR ADDITION SYSTEMS

Word problems constitute a very natural, as well as an old, widely studied, class of decision problems. The basic idea is the following. We begin with finitely many equations

$$w_1 = w_1', \ldots, w_k = w_k', \tag{11}$$

where the w's are words. We want to find an algorithm for deciding whether or not an arbitrary given word w can be transformed into another given word w' using equations (11). Without any formal definition, it should be clear what *using equations* means: We are allowed to substitute a subword w_i by w_i', or vice versa.

The problem formulated above is the **word problem for semigroups**. Equations (11) define a particular finitely generated semigroup. The algorithm for which we are looking would decide for an arbitrary pair (w, w'), where w and w' are words over the alphabet of generators, whether or not $w = w'$ holds in the semigroup. The **word problem for groups** is defined similarly. Of course, it is a special case of the word problem for semigroups because, in case of groups, among equations (11) there are always particular equations stating the properties of the inverses and the identity.

Word problems can be stated for (finitely generated) **Abelian** (commutative) groups and semigroups as well. Altogether, this gives us four algorithmic problems concerning the most fundamental algebraic structures.

Word problems are customarily referred to also as **Thue problems**. Indeed, Thue considered rewriting systems with symmetric productions $w \leftrightarrow w'$, which amounts to the study of equations used as defining relations. By a **semi-Thue problem** we mean the problem where the productions are ordinary (as in case of rewriting systems) nonsymmetric ones: $w \to w'$. In this fashion, we obtain eight very natural algorithmic problems for fundamental algebraic structures: the Thue and semi-Thue problem for groups and semigroups, as well as for Abelian groups and Abelian semigroups.

It turns out that in this setup the borderline between decidability and undecidability can be drawn: All four problems are decidable for Abelian structures, whereas all four are undecidable for arbitrary structures. To obtain this result, it suffices to show the undecidability (resp. decidability) of the smallest (resp. largest) problem—that is, the Thue problem for groups (resp. the semi-Thue problem for Abelian semigroups). Historically, the development took a different line. For instance, it is much easier to prove the undecidability of the Thue problem for semigroups than for groups. Indeed, this easier proof was given some 10 years earlier than the more-difficult one. A variant of this proof is also given below.

Theorem 5.13. *The Thue problem for semigroups is undecidable. More specifically, there is a semigroup S with finitely many defining relations $w_i = w_i'$, $i = 1, \ldots, k$, such that there is no algorithm for deciding whether or not an arbitrary equation $w = w'$ holds in S.*

Proof. We modify the definition of Turing machines presented in Section 4.1 in the following way. For every letter a in the operating alphabet Σ_O, the productions

$$aq_F \rightarrow q_F \quad \text{and} \quad q_F a \rightarrow q_F \tag{12}$$

are added to the set P of productions. Consequently, a Turing machine halts with a word w iff

$$\#q_0 w\# \Rightarrow {}^* q_F.$$

It is obvious that this change concerning the final state q_F in no way affects the computing capabilities of Turing machines. In particular, Lemma 5.3 remains valid: There is an index i_0 for which the function $g(x)$ defined in Lemma 5.3 is not recursive. (Thus, the Turing machine TM_{i_0} now has the additional productions in (12).)

Consider the semigroup S, whose defining relations are obtained by transforming the productions of TM_{i_0} into equations. Thus, $w_i = w_i'$ is a defining relation of S iff $w_i \rightarrow w_i'$ is a production of TM_{i_0}.

We know that for an arbitrary word w over the terminal alphabet Σ_T of TM_{i_0}, the machine TM_{i_0} halts with w iff

$$\#q_0 w\# \Rightarrow {}^* q_F \tag{13}$$

holds according to the rewriting system determined by TM_{i_0}. Hence, there is no algorithm for deciding whether or not (13) holds for an arbitrary w.

We now claim that (13) holds iff

$$\#q_0 w\# = q_F \tag{14}$$

is valid in S. Clearly, this claim implies Theorem 5.13 because deciding the validity of equations (14) constitutes only a special case of the word problem for S.

To establish our claim, we first observe that whenever (13) holds, (14) also holds. This follows because we can obviously replace an application of any production in TM_{i_0} by an application of the corresponding equation in S.

Conversely, assume that (14) holds. Consequently, there is a finite sequence of words

$$x_0 = \#q_0w\#, x_1, \ldots, x_n = q_F \tag{15}$$

such that every x_i, $1 \le i \le n$, is obtained from x_{i-1} by using some of the defining relations of S. If there are distinct numbers i and j such that $x_i = x_j$ in (15), we may replace (15) by a shorter sequence of words satisfying the same conditions. Hence, we may assume without loss of generality that all words x_i in (15) are distinct.

Let p be the smallest number such that in (15) the word x_p contains an occurrence of q_F. Using productions (12), we see that

$$x_p \Rightarrow^* q_F \tag{16}$$

holds for TM_{i_0}. Of course, the subsequence $x_p, \ldots, x_n$ of (15) may originally contain derivation steps to which no computation step of TM_{i_0} corresponds because the equations may have been used in (15) in the "wrong direction." (Indeed, this subsequence of (15) may even contain words without an occurrence of q_F if one has used an equation resulting from a production introducing q_F in the wrong direction.) However, this is irrelevant because the occurrence of q_F in x_p opens the possibility of applying productions (12), which shows that (16) holds.

To establish (13), we still have to show that

$$x_0 \Rightarrow^* x_p \tag{17}$$

holds for TM_{i_0}. By the choice of p, none of the words $x_0, \ldots, x_{p-1}$ contains an occurrence of q_F. We now study the equations used to obtain the subsequence $x_0, \ldots, x_p$ of (15). In this subsequence, if every word x_i with $i \ge 1$ results from the previous one, x_{i-1}, by replacing the left side of some production of TM_{i_0} by its right side, we conclude that (17) holds. Thus, assume that this is not the case—that is, that some equation has been used in the wrong direction at least once. Let $r \le p$ be the greatest index such that x_r results in (15) from x_{r-1} by replacing the right side of some production of TM_{i_0} by its left side. Clearly, $r \le p - 1$ because q_F is introduced at the step from x_{p-1} to x_p and, hence, this step uses a production in the correct direction.

Thus, we know that

$$x_{r-1} = y_1\beta y_2, \qquad x_r = y_1\alpha y_2$$

for some production $\alpha \to \beta$ of TM_{i_0} and for some words y_1 and y_2. It is now easy to see that $\alpha \to \beta$ is the only production applicable to x_r. Indeed, the state symbol determines the position in a word where a production can be applied, and checking through the different productions (1)–(5) in the definition of a Turing machine given in Section 4.1, we conclude that $\alpha \to \beta$ is the only possibility. (This is basically due to the deterministic mode of operation of Turing machines.)

On the other hand, our assumption concerning r guarantees that x_{r+1} results from x_r by an application of some production of TM_{i_0} (in the correct direction). Consequently

$$x_{r+1} = y_1 \beta y_2 = x_{r-1},$$

which contradicts the fact that all words in (15) are distinct. Hence, we conclude that (17) and thus also (13) holds.   □

To obtain the semigroup S of Theorem 5.13 explicitly, we have to list the productions of TM_{i_0}, which—of course—can be done. A simpler example of a semigroup S satisfying Theorem 5.13 is given in Exercise 5.6.

Theorem 5.13 implies that also the semi-Thue problem for semigroups is undecidable. The proof of the undecidability of the Thue problem for groups lies beyond the scope of this book. The proof of the decidability of the Thue problem for Abelian groups and semigroups belongs essentially to linear algebra and should present no difficulties for an interested reader. On the other hand, the decidability of the semi-Thue problem for Abelian semigroups was established only recently, in the early 1980s. Thus, among the word problems we have been considering, this was open for the longest time. While the proof of this result is omitted, we shall develop another very natural formulation for the underlying decision problem. Indeed, the problem can be viewed from many different angles—still another formulation is given in Exercise 5.5.

Let $n \geq 2$ be an integer. An *n*-dimensional vector addition system is a finite set

$$VAS = \{v_1, \ldots, v_k\},$$

where each v_i is an ordered *n*-tuple consisting of integers. Such *n*-tuples are referred to as **vectors**. Let u and u' be vectors with nonnegative components. Then u' is **reachable** from u according to VAS iff there is a finite sequence

$$u_0 = u, u_1, \ldots, u_m = u'$$

of vectors with *nonnegative components* such that for each i, $1 \leq i \leq m$, there is a j, $1 \leq j \leq k$, with the property

$$u_i = u_{i-1} + v_j.$$

(The addition is carried out componentwise.)

For instance, if $n = 2$ and

$$VAS = \{(1, 0), (0, 1), (-100, -100)\}$$

then for any u and u', u' is reachable from u according to VAS.

The **reachability problem for vector addition systems** consists of finding an algorithm that will decide of an arbitrary triple (VAS, u, u') whether or not u' is reachable from u according to VAS.

The key issue in the definition of reachability is that no negative components are allowed in the intermediate steps. Otherwise, we are essentially dealing with a problem in linear algebra. A much-easier version of the reachability problem is given in Exercise 5.7.

Theorem 5.14. *The reachability problem for vector addition systems and the semi-Thue problem for Abelian semigroups are equivalent with respect to decidability: An algorithm for solving one of the problems can be converted to an algorithm for solving the other.*

Proof. Observe first that since we are dealing with Abelian semigroups, we may assume that the generators $a_1, \ldots, a_n$ appear in alphabetic order in all words. For a word $w = a_1^{i_1} \cdots a_n^{i_n}$ where $i_j \geq 0$, we let

$$V(w) = (i_1, \ldots, i_n)$$

and for a vector $v = (i_1, \ldots, i_n)$ with nonnegative components, we let

$$W(v) = a_1^{i_1} \cdots a_n^{i_n}.$$

These notations allow us to speak of vectors associated to words and of words associated to vectors.

Assume first that an algorithm for the semi-Thue problem for Abelian semigroups is known. Then the reachability problem can be solved as follows. Given a triple (VAS, u, u'), we decide whether or not

$$W(u) \Rightarrow_S^* W(u') \tag{18}$$

holds in the Abelian semigroup S with generators $a_1, \ldots, a_n$, where the productions are defined as follows. Consider a vector v in VAS. Let v' be the vector obtained from v by replacing all positive components with 0 and changing all negative components x into $-x$. Similarly, let v'' be the vector contained from v by replacing all negative components with 0. Then $W(v') \to W(v'')$ is a production. All productions are obtained in this fashion. (The empty word λ may occur in the productions. It is then an identity element for S. This property has to be stated in the formal description of S.)

It is clear that the productions of S simulate the addition of vectors v in VAS. A production will not be applicable if the addition of the corresponding vector would introduce a negative component. Hence, u' is reachable from u according to VAS iff (18) holds.

Conversely, assume that an algorithm for the reachability problem is known. Then the semi-Thue problem for Abelian semigroups can be solved as follows. Given an Abelian semigroup S and two words w and w' over the alphabet $\{a_1, \ldots, a_n\}$ of generators of S, we want to decide whether or not

$$w \Rightarrow_S^* w' \tag{19}$$

holds.

Let

$$w_i \to w_i', \qquad i = 1, \ldots, k, \tag{20}$$

be the productions of S. Consider the $(n + 3)$-dimensional vector addition system VAS, consisting of the vectors

$$(-V(w_i), -1, i, 2k + 1 - i) \quad \text{and} \quad (V(w_i'), 1, -i, i - 2k - 1) \tag{21}$$

for $i = 1, \ldots, k$. (The notation $-V(w_i)$ stands for the vector obtained from $V(w_i)$ by replacing every component x with $-x$.)

It is now easy to see that (19) holds iff the vector $u' = (V(w'), 1, 0, 0)$ is reachable from the vector $u = (V(w), 1, 0, 0)$ according to VAS. In fact, the three last components in a permissible computation are always

$$(1, 0, 0) \quad \text{or} \quad (0, i, 2k + 1 - i), \tag{22}$$

for some $i = 1, \ldots, k$. Whenever the first (resp. second) of the alternatives in (22) holds, we must add, for some i, the first (resp. second) of the vectors in (21) in order to avoid negative components, after which the second (resp. first) of the alternatives (22) holds. The first of the vectors in (21) removes the letters on the left side of some production (20). The value of i is remembered by the last two components of the resulting vector, and only the correct choice of the second vector from (21) at the next step avoids the introduction of negative components.   □

The second part of the proof of the previous theorem shows in a special case how vector addition systems with "states" can be simulated by ordinary vector addition systems. Further details can be found in Exercise 5.8.

We summarize, without proof, the decidability status of the different word problems.

Theorem 5.15. *The Thue problem (and, consequently, the semi-Thue problem) for groups is undecidable. The Thue and semi-Thue problems for Abelian groups and semigroups are decidable.*

EXERCISES

5.1. Consider the construction of Theorem 5.1 for the grammar G presented in Example 2.5. Show how the construction works for the terminal words $w = a^2b^2c^2$ and $w = a^2b^2c^3$.

5.2.   Prove the undecidability of the following problems. (a) Is a given context-free language regular? (b) Is a given context-free language equal
to a given regular language? (c) Is the complement of a given context-
free language empty? (d) Is the intersection of two given context-free
languages finite? State and study similar problems for 0L languages.
Observe that the family of D0L languages has strong decidability
properties. For instance, the equivalence problem is decidable.

5.3.   In the definition of the *modified* Post correspondence problem, only
words beginning with a specified letter a are accepted as solutions.
Show that also the modified Post correspondence problem is undecidable.

5.4.   Give an upper bound for the degree of the polynomial Q in Theorem
5.7.

5.5.   A **matrix grammar** consists of finitely many finite sequences
$[A_1 \rightarrow \alpha_1, \ldots, A_k \rightarrow \alpha_k]$ (matrices) of context-free productions. An application of such a matrix to a word consists of replacing an occurrence
of A_1 by α_1 and so forth until, finally, an occurrence of A_k is replaced
by α_k. If some production in a matrix is not applicable when its turn
comes, the derivation halts without producing a terminal word. Nonterminals and terminals, as well as the start letter, are specified in the
same way as for ordinary context-free grammars. The details can be
found in [Sa3]. Prove that the reachability problem for vector-addition systems and the emptiness problem for languages generated by
matrix grammars are equivalent with respect to decidability.

5.6.   Prove that the Thue problem is undecidable for the semigroup S with
the generators a, b, c, d, e and defining relations

$$ac = ca, \qquad ad = da, \qquad bc = cb, \qquad bd = db,$$
$$aba = abae = eaba, \qquad eca = ae, \qquad edb = be.$$

(*Hint:* One uses the fact that the equation $w = \lambda$ is undecidable for
monoids M with defining relations

$$w_1 = \lambda, \; w_2 = \lambda, \ldots, w_n = \lambda. \tag{23}$$

Two encodings are used for the letters a_i of an arbitrary M:

$$\varphi(a_i) = ab^i a \quad \text{and} \quad \psi(a_i) = cd^i c.$$

Since neither of the pairs (a, b) and (c, d) commutes, the information
thus encoded is preserved in S. One now asks the question: Does
$\varphi(M)\psi(w) = \varphi(M)$ hold in S? Here $\varphi(M)$ gives the encoding of the relations in (23), separated by some boundary marker. This question is
equivalent to the question: Does $w = \lambda$ hold in M? The reason for this
is that the letter e controls erasing in S, according to the defining relations of M.)

5.7. Show that the following problem is decidable. Given VAS = $\{v_1, \ldots, v_k\}$ and u, u', do there exist nonnegative integers α_i with the property

$$u + \alpha_1 v_1 + \cdots + \alpha_k v_k = u'?$$

5.8. Consider vector addition systems with **states,** defined as follows. Together with VAS, a finite directed graph G with some specified initial and final nodes is given. The arrows of G are labeled with vectors of VAS. The graph G controls the addition process in an obvious fashion: At each step the node visited determines a subset of vectors that can be used, and the vector actually chosen then determines the next node. The reachability question is in this setup: given u and u', can one start with u from an initial node and end with u' in a final node? Show that n-dimensional vector-addition systems with states can be simulated with ordinary $(n + 3)$-dimensional vector-addition systems. (The result is due to [HoP].) Consequently, the decidability of the reachability problem for ordinary vector addition systems implies the decidability of the same problem for systems with states.

5.9. Consider the following decision problems. Given an n-dimensional square matrix M with integer entries. (a) Decide whether or not there exists a positive integer n such that the number 0 appears in the upper right-hand corner of M^n. (b) Decide whether or not, for all positive integers n, the number in the upper right-hand corner of M^n is greater than or equal 0. Problems (a) and (b) are among the most famous ones whose decidability status is open. They are also very significant ones because quite a few open problems in automata and language theory have been reduced to these problems. Further information can be found in [SaS] and [RozS].

Show that, in regard to decidability, problem (a) is equivalent to the problem of deciding, given a D0L growth function f, whether or not the equation $f(n) = f(n + 1)$ holds for some integer n. Similarly, show that problem (b) is equivalent to the problem of deciding whether or not a given D0L growth function is monotonic.

CHAPTER 6

Computational Complexity

6.1. BASIC IDEAS AND AXIOMATIC THEORY

The classification of mathematical problems into decidable and undecidable ones has been discussed quite extensively in previous chapters. Indeed, from the mathematical point of view, this classification is the most fundamental one. It is also of definite practical significance in discouraging attempts to design too-general systems, such as systems for deciding the equivalence of programs or the halting of a program. There are, in fact, instances of such attempts in the past!

However, this classification is too coarse in many respects. In Chapter 4 we considered a finer classification of *undecidable* problems in terms of reducibilities and degrees. We shall now discuss a finer classification of *decidable* problems in terms of their complexity. Two problems might both be decidable and yet one might be enormously more difficult to compute, which in practice might make this problem resemble an undecidable one. For example, if instances of reasonable size of one problem take only a few seconds to compute, whereas instances of comparable size of the other problem take millions of years (even if best computers are available), it is clear that the latter problem should be considered intractable compared with the former one. Hence, having established the decidability of a particular problem, we should definitely study its complexity—that is, how difficult it is to settle specific instances of the problem. Complexity considerations are of crucial importance, for instance, in cryptography. If we are not able to decrypt a message within a certain time limit, we might as well forget the whole thing because, as time passes by, the situation might change entirely.

Thus, typical questions in the theory of computational complexity would be: Is the algorithm A_1 better than the algorithm A_2 (for the same problem) in the sense that it uses fewer resources, such as time or memory space? Does the problem P have a best algorithm? Is the problem P more difficult than the problem P' in the sense that every algorithm for solving P is more complex than some fixed reasonable algorithm for solving P'?

Natural complexity measures are obtained by considering Turing machines: How many steps (in terms of the length of the input) does a computation require? This is a natural formalization of the notion of time resource. Similarly, the number of squares visited during a computation constitutes a natural space measure.

We shall return to such *machine-oriented* complexity considerations in the second half of this chapter. Our discussions will be restricted to the basic Turing machine model. Variations such as machines with many tapes and many heads or random access machines are important in more detailed and specific complexity considerations. However, from the point of view of the most fundamental issues, the particular Turing machine model chosen is irrelevant.

In the first half of this chapter, we shall study **axiomatic complexity** theory. No specific complexity measure, such as time or memory space, will be defined. Instead, we just speak of an "abstract resource" used by an algorithm. The axioms applied are very natural. They also look very weak in the sense that they do not say much. However, quite a remarkable theory based on the axioms can be established. Moreover, this theory is a natural sideline of the theory of partial recursive functions developed in Chapter 4. Of course, the weakness of the axioms is somehow reflected in the results: some "pathological" results can be established.

We do not discuss at all the third major aspect of complexity theory: *low-level complexity,* or the complexity of some specific but practically important algorithms and problems. The reader is referred to [AHU], [GaJ], and [MaY] for this topic, as well as for more-detailed information of the broad and highly developed area of complexity theory in general.

We now begin the study of axiomatic complexity theory. Although abstract, it gives us definite information about specific machine-oriented measures such as computing time or memory space. Sometimes proofs are even simpler in the axiomatic theory than in a theory dealing with a specific measure.

Consider again an enumeration

$$TM_1, TM_2, \ldots, TM_i, \ldots \tag{1}$$

of all Turing machines, satisfying the same requirements as the enumeration (6) in Chapter 4. In other words, assume that for (1) there are effective procedures for going from the index to the machine and from the machine to the index. Enumeration (1) determines an enumeration

$$\bar{f}_1, \bar{f}_2, \ldots, \bar{f}_i, \ldots \tag{2}$$

of all partial recursive functions of *one* variable. An enumeration

$$f_1, f_2, \ldots, f_i, \ldots \tag{3}$$

of all partial recursive functions of one variable is termed **acceptable** iff it is possible to go effectively from (2) to (3), and vice versa. In other words, given an index for a function in (2), we can find an index for the same function in (3), and vice versa. More information about acceptable enumerations is contained in Exercise 6.1.

> **Definition 6.1.** *A complexity measure is a pair* $CM = (F, \Phi)$, *where* F *is an acceptable enumeration* (3) *of partial recursive functions and* Φ *is an infinite sequence*
>
> $$\varphi_1, \varphi_2, \ldots, \varphi_i, \ldots \tag{4}$$
>
> *of partial recursive functions such that the* **Blum axioms** *B1 and B2 are satisfied.*
>
> **B1.** *For each* $i \geq 1$, *the domains of* f_i *and* φ_i *coincide.*
> **B2.** *The function* $g(i, n, m)$ *defined by*
>
> $$g(i, n, m) = \begin{cases} 1 & \text{if } \varphi_i(n) = m \\ 0 & \text{otherwise} \end{cases}$$
>
> *is recursive.*
>
> *The function* φ_i *is referred to as the* **complexity function**, *or cost function*, *of* f_i. $\qquad\square$

For instance, if we choose $\varphi_i(n)$ to be the number of steps in the computation of a Turing machine for f_i for the input n, we clearly obtain a complexity measure. Similarly, a complexity measure results if we let $\varphi_i(n)$ be the number of squares visited during the computation of the same Turing machine for the input n, provided such a variant of Turing machines is considered, where no machine loops using only a finite amount of tape. Further, similar complexity measures are mentioned in Exercise 6.2.

On the other hand, the choice $\varphi_i = f_i$ does not yield a complexity measure: Axiom B2 is not satisfied. It would contradict, for instance, Theorem 4.7. The choice $\varphi_i(n) = 1$ for all i and n does not yield a complexity measure because Axiom B1 is not satisfied. These examples also show that the two axioms are independent.

Clearly, every partial recursive function f occurs infinitely many times in sequence (3). If $f = f_i$, then the cost function φ_i is associated to an algorithm (for instance, a Turing machine) for computing f, rather than to the function f itself. If $f = f_j$ as well, then the cost function φ_j might have essentially smaller values than φ_i, showing that the algorithm corresponding to f_j is essentially better. When and how is such a speedup possible? This question will be discussed in Section 6.3.

In general, the complexity measures introduced above are **machine-independent** in the sense that it is not specified what machines or algorithms correspond to f_i.

It is an immediate consequence of Axiom B2 that all the functions

$$g_1(i, n, m) = \begin{cases} 1 & \text{if } \varphi_i(n) \leq m \\ 0 & \text{otherwise} \end{cases}$$

and

$$g_h(i, n, m) = \begin{cases} 1 & \text{if } \varphi_i(n) \leq h(m) \\ 0 & \text{otherwise,} \end{cases}$$

where h is an arbitrary recursive function, are recursive. This fact will be used in the sequel without further mention.

It was already pointed out above that because of the generality of Axioms B1 and B2, some pathological complexity measures will be obtained. For instance, we may compute our favorite recursive functions without any cost whatsoever. Suppose we want to test the primality of integers; that is, our favorite function is the characteristic function $h_P(n)$ for the set of prime numbers. Let j be an index for h_P. Consider an arbitrary sequence φ_i of cost functions yielding a complexity measure. Define a sequence φ_i' by

$$\varphi_i'(n) = \begin{cases} \varphi_i(n) & \text{for } i \neq j \\ 0 & \text{for } i = j. \end{cases}$$

Clearly, φ_i' also yields a complexity measure, and we have a free primality test according to this measure!

In spite of such pathological features, quite an extensive theory of complexity measures can be developed. In Sections 6.1–6.3 we present the basics of such a theory. The present section deals mainly with results showing how complex algorithms can be according to different measures. We begin with a rather pleasing result to the effect that we can find out an index for any cost function.

Theorem 6.1. *For every complexity measure* $CM = (F, \Phi)$, *there is a recursive function* h *such that* $\varphi_i = f_{h(i)}$ *holds for every* i.

Proof. Define a function $f(x, i)$ of two variables by

$$f(x, i) = \varphi_i(x).$$

Clearly, by Axiom B2, f is partial recursive. Let k be an index for f. By the s-m-n theorem (Theorem 4.1), there is a recursive function s such that

$$f(x, i) = f_{s(k, i)}(x).$$

(The s-m-n theorem holds by Exercise 6.1 for the acceptable enumeration (3) as well.) Theorem 6.1 now follows by choosing $h(i) = s(k, i)$. □

Our next theorem exhibits an intuitively obvious fact: A bad program may use an arbitrarily large amount of resource even for the computation of a simple function. In the statement of this theorem, as well as the following theorems, we have an arbitrary complexity measure in mind. In connection with the measure, we use the notations specified in Definition 6.1. Moreover, we use the notation IND(f) to mean the set of all indices of f in the acceptable enumeration (3).

Theorem 6.2. *Let f and h be recursive functions. Then, for some $j \in IND(f)$,*

$$\varphi_j(x) > h(x)$$

holds for every x.

Proof. Consider the function ψ defined by

$$\psi(x, i) = \begin{cases} f(x) & \text{if } \varphi_i(x) \le h(x) \text{ is not true} \\ f_i(x) + 1 & \text{otherwise.} \end{cases}$$

The function $\psi(x, i)$ is (total) recursive because if the first line in the definition is not satisfied (that is, $\varphi_i(x) \le h(x)$) then we have $f_i(x)\downarrow$ by Axiom B1. By the s-n-m theorem, there is a recursive function s such that

$$\psi(x, i) = f_{s(i)}(x). \tag{5}$$

An application of the recursion theorem (Theorem 4.6; also see Exercise 6.3) for the function s yields an index j such that

$$f_{s(j)} = f_j. \tag{6}$$

Equations (5) and (6) give the result

$$\psi(x, j) = f_j(x).$$

This implies, by the definition of ψ, that the equation $f_j(x) = f(x)$ holds for every x. Hence, $j \in IND(f)$. Since f is recursive, $\varphi_j(x)\downarrow$ for every x. Moreover, $\varphi_j(x) > h(x)$ holds for every x because the first line of the definition is valid for $\psi(x, j)$ independently of x. □

While the result of Theorem 6.2 might be due to bad programming, the next result opens a more-pessimistic view: Even excellent programs might use indefinite amounts of resources. This follows because, according to our next theorem, there are arbitrarily complex functions.

Theorem 6.3. *Let $f(x)$ be an arbitrary recursive function. Then there is a recursive function h such that for every j in $IND(h)$, there are infinitely many values of x with the property $\varphi_j(x) > f(x)$.*

Proof. First consider the recursive function $\alpha(x)$, assuming the values

$$001012012301234012345\ldots$$

(in this order) for the argument values $x = 0, 1, 2, \ldots$. Then, for every $y \geq 0$, there are infinitely many values of x such that $\alpha(x) = y$.

Our function h is now defined by

$$h(x) = \begin{cases} f_{\alpha(x)}(x) + 1 & \text{if } \varphi_{\alpha(x)}(x) \leq f(x) \\ 0 & \text{otherwise.} \end{cases}$$

By Axiom B2, the condition $\varphi_{\alpha(x)}(x) \leq f(x)$ can be tested effectively and, consequently, h is a recursive function. Indeed, h is total recursive because, by Axiom B1, whenever the first line of the definition is applicable, then $f_{\alpha(x)}(x)\downarrow$.

Assume now that j is in $IND(h)$. By the definition of α, there are infinitely many values of x satisfying $\alpha(x) = j$. For each of these numbers x, we obtain

$$\varphi_{\alpha(x)}(x) > f(x), \quad \text{that is,} \quad \varphi_j(x) > f(x).$$

This is a direct consequence of the definition of h. $\qquad\square$

Theorem 6.3 was obtained by a diagonalization argument. A more-complicated diagonalization yields the following stronger result. The result shows that there are recursive functions assuming only the values 0 and 1, which are arbitrarily complex in the sense that, from a certain bound on, their complexity exceeds any pregiven recursive function. We use the expression *almost everywhere* to mean that the condition in question holds from a certain bound on—that is, it holds universally with finitely many exceptions. (This condition is clearly stronger than the condition *infinitely often* discussed in Theorem 6.3.)

Theorem 6.4. *There is a recursive function h such that for every recursive function f, there exists a recursive function e satisfying conditions (i)–(iii).*

  (i) *Every i in $IND(f)$ satisfies $f_{h(i)} = e$.*
  (ii) *Every i in $IND(e)$ satisfies the condition*

$$\varphi_i(x) > f(x)$$

almost everywhere.

(iii) *e assumes only the values* 0 *and* 1.

Proof.   Given f, we simultaneously define two functions s and e as follows.

$$s(0) = \begin{cases} 0 & \text{if } \varphi_0(0) \le f(0) \\ \uparrow & \text{otherwise;} \end{cases}$$

$$s(x+1) = \begin{cases} \min_{k \le x+1} & [\varphi_k(x+1) \le f(x+1), \text{ and there is} \\ & \text{no } y \le x \text{ such that } \varphi_k(y) \le f(y) \\ & \text{and } f_k(y) \ne e(y)] \\ \uparrow & \text{if no such } k \text{ exists;} \end{cases}$$

$$e(x) = \begin{cases} 0 & \text{if } f_{s(x)}(x) = 1 \\ 1 & \text{otherwise.} \end{cases}$$

The inequalities in the definition of s are decidable, by Axiom B2. Moreover, the condition $s(x)\!\downarrow$ is decidable and, whenever $s(x)\!\downarrow$, then $f_{s(x)}(x)\!\downarrow$, which implies that the equation $f_{s(x)}(x) = 1$ is decidable. (If $s(x)\!\uparrow$, then the second line of the definition of $e(x)$ applies.) Consequently, s is partial recursive and e is recursive.

The definition of e guarantees that condition (iii) is satisfied. Consider condition (ii). We claim first that the function s is injective. Assume the contrary: There are numbers x_1 and x_2, $x_1 < x_2$, such that $s(x_1) = s(x_2) = k$. By the definition of $s(x_2)$, we infer that $f_k(x_1) = e(x_1)$. This, however, contradicts the definition of $e(x_1)$, and hence our claim follows.

Assume now that condition (ii) is not satisfied. This means that for some i in IND(e), there are infinitely many numbers x with the property

$$\varphi_i(x) \le f(x). \tag{7}$$

Consequently, there are infinitely many numbers x with the property (mentioned in the definition of $s(x+1)$)

$$\varphi_i(x) \le f(x), \quad \text{and there is no } y < x \text{ such that}$$
$$\varphi_i(y) \le f(y) \quad \text{and} \quad f_i(y) \ne e(y). \tag{8}$$

Indeed, (8) follows from (7) because the additional condition is always satisfied: $f_i(y) \ne e(y)$ holds for no y because i is an index for e.

Since s is injective, there is a number x_0 such that no number less than or equal i appears as a value of s for argument values greater than x_0. On the other hand, because (8) holds for infinitely many numbers x, there is a number $x' > x_0$ satisfying (8). According to the definition of s, we have $s(x') \le i$, which contradicts the fact that no number less than or equal i appears as a value of s for argument values greater than x_0. Hence, condition (ii) is satisfied.

Finally, condition (i) is an immediate consequence of the fact that the construction of e is uniform effective in terms of f. $\square$

As an immediate consequence of Theorem 6.4, we get the result that there is no recursive bound for the values of the cost function in terms of the values of the corresponding partial recursive function.

Theorem 6.5.   *There is no recursive function α such that for every i,*

$$\varphi_i(x) \le \alpha(x, f_i(x)) \tag{9}$$

holds almost everywhere.

Proof.   Assume the contrary: Such a recursive function α exists. Then the function

$$\alpha_1(x) = \alpha(x, 0) + \alpha(x, 1) \tag{10}$$

is also recursive. We now choose $\alpha_1 = f$ in Theorem 6.4. Consequently, there is a recursive function e assuming only the values 0 and 1 and such that $\varphi_i(x) > \alpha_1(x)$ holds almost everywhere for all i in IND(e). For these values of i, we now have a contradiction with (9) because of (10) and the fact that for all x, $f_i(x) = 0$ or $f_i(x) = 1$.   □

On the other hand, there is a recursive bound for the values of a partial recursive function in terms of the values of the corresponding cost function.

Theorem 6.6.   *There is a recursive function α such that for every i,*

$$f_i(x) \le \alpha(x, \varphi_i(x)) \tag{11}$$

holds almost everywhere.

Proof.   The function ψ defined by

$$\psi(i, x, y) = \begin{cases} f_i(x) & \text{if } \varphi_i(x) = y \\ 0 & \text{otherwise} \end{cases}$$

is clearly recursive, by Axiom B2. (In fact, Axiom B1 is also needed because if $\varphi_i(x) = y$, then $f_i(x)\downarrow$ according to B1.)
     Now define

$$\alpha(x, y) = \max_{i \le x} \{\psi(i, x, y)\}.$$

Hence, if $x \ge i$, then (11) holds.   □

Having studied bounds for cost functions in terms of partial recursive functions and the other way around, we now study bounds for cost functions in terms of corresponding cost functions in another measure. The next theorem shows that two measures cannot be too different in this respect, which is another pleasing result. The result is customarily referred to as the theorem about **recursive relatedness**.

Theorem 6.7.   *Let the cost functions φ_i and φ_i' both define a complexity measure for the sequence f_i of partial recursive functions. Then there exists a recursive function h such that for every i, the equations*

$$\varphi_i'(x) \leq h(x, \varphi_i(x)) \quad and \quad \varphi_i(x) \leq h(x, \varphi_i'(x)) \tag{12}$$

hold almost everywhere.

Proof.   Define

$$h_1(i, x, y) = \begin{cases} \varphi_i(x) + \varphi_i'(x) & \text{if } \varphi_i(x) \leq y \text{ or } \varphi_i'(x) \leq y \\ 0 & \text{otherwise.} \end{cases}$$

Given an argument value (i, x, y), we can immediately compute $h_1(i, x, y)$ by Axiom B2. (Observe also that if either $\varphi_i(x)$ or $\varphi_i'(x)$ is less than or equal y, then Axiom B1 guarantees that both are defined, and so the sum can be effectively computed.) Hence, h_1 is recursive, and so is the function h defined by

$$h(x, y) = \max_{i \leq x} h_1(i, x, y).$$

This definition guarantees that (12) holds for every $x \geq i$. Consequently, (12) holds almost everywhere in x. □

Theorem 6.7 expresses the fact that any function easy to compute according to some measure is easy to compute according to all measures.

6.2.   COMPLEXITY CLASSES, GAP, AND COMPRESSION THEOREMS

A recursive function α determines a class C_α of recursive functions in the following way. A function is in C_α iff it has an algorithm whose complexity is no greater than α. Such classes are referred to as *complexity classes*. The following formal definition is again based on a fixed underlying complexity measure. Of course, the classes C_α depend on this measure. If there is some danger of confusion, we indicate the measure in the notation by writing C_α^φ instead of C_α.

Definition 6.2.   *The complexity class of a recursive function α is defined by*

$$C_\alpha = \{recursive \ f \mid f = f_i, \ for \ some \ i \ such \ that \ \varphi_i(x) \leq \alpha(x)$$
$$almost \ everywhere\}. \quad □$$

The existence of arbitrarily complex recursive functions (for instance, Theorem 6.4) guarantees that for any complexity class C_α, there are recursive functions not belonging to C_α. This implies that for every α, there is an α_1 such that $C_\alpha \subsetneq C_{\alpha_1}$. This yields an infinite ascending chain of complexity classes

$$C_\alpha \subsetneq C_{\alpha_1} \subsetneq C_{\alpha_2} \subsetneq \cdots.$$

On the other hand, if $\alpha(x) \leq \beta(x)$ almost everywhere, then $C_\alpha \subseteq C_\beta$. When is the inclusion strict? The following **gap theorem** shows that even arbitrarily big increases in α—for instance the choice

$$\beta(x) = 2^{2^{\alpha(x)}}$$

—do not guarantee that the inclusion $C_\alpha \subseteq C_\beta$ is strict. Thus, the interval between α and β might constitute a gap in the sense that there are no algorithms whose complexity falls between α and β. This result also shows that there cannot be any uniform effective procedure for finding a recursive function α_1 given a recursive function α such that $C_\alpha \subsetneq C_{\alpha_1}$. In the statement of the theorem, the notation $h\alpha$ stands for composition $h\alpha(x) = h(\alpha(x))$ for all x.

Theorem 6.8. *Let h be a recursive function satisfying $h(x) \geq x$ for all x. Then there are arbitrarily large recursive functions α with the property*

$$C_\alpha = C_{h\alpha}. \tag{13}$$

Proof. Consider the function α defined by

$$\alpha(0) \quad = 1,$$
$$\alpha(x + 1) = \alpha(x) + m_x,$$

where m_x is the least integer $m \geq 1$ such that for each $i \leq x + 1$, either $\varphi_i(x + 1) \leq \alpha(x) + m$ or else it is not the case that

$$\varphi_i(x + 1) \leq h(\alpha(x) + m). \tag{14}$$

We first note that m_x always exists, since if $\varphi_i(x + 1)$ is not defined, then (14) is not satisfied for any m. We are able to compute m_x by Axiom B2 and, hence, we conclude that α is recursive.

The definition of α guarantees that for all j, whenever

$$\varphi_j(x) \leq h(\alpha(x))$$

holds almost everywhere, then

$$\varphi_j(x) \leq \alpha(x)$$

also holds almost everywhere. This implies that (13) is satisfied. We may replace α by an arbitrarily large function simply by changing the initial condition $\alpha(0) = 1$ and adding to the definition of m_x a condition to the effect that $\alpha(x + 1)$ will exceed the considered (recursive) lower bound. $\square$

The gap theorem asserts that for any complexity measure, there are arbitrarily large complexity gaps—that is, intervals in which no cost functions can be found. No recursive function h has the property that $C_\alpha \subsetneq C_{h\alpha}$ holds for all recursive functions α. We can interpret this to mean that cost functions are sparse with respect to recursive functions. Also, the following rather provocative interpretation can be given.

Consider two computers, FOSSIL and FUTURA. FOSSIL is old-fashioned and really slow, whereas FUTURA operates at fantastic speeds and has all the features of which you might dream. We still claim that there are time bounds β (arbitrarily large) such that every algorithm realized by FUTURA within time bound β is realized also by FOSSIL within time bound β. Thus, within such time bounds, FOSSIL is able to catch FUTURA!

Indeed, let φ_i and φ_i' be the cost functions determined by the computing times of FOSSIL and FUTURA, respectively. (We assume that FOSSIL is able to simulate Turing machines, so that partial recursive functions can be computed.) Although in general, $\varphi_i(x)$ is enormous in comparison with $\varphi_i'(x)$, there is a recursive function h such that

$$\varphi_i(x) \le h(\varphi_i'(x)) \tag{15}$$

holds for all i and x. (See also Theorem 6.7.) We now choose, according to Theorem 6.8, a recursive function α such that $C_\alpha^{\varphi'} = C_{h\alpha}^{\varphi'}$. Now assume a function f_j belongs to $C_{h\alpha}^{\varphi'}$—that is, $\varphi_j'(x) \le h\alpha(x)$ almost everywhere. (More explicitly, a function f is in $C_{h\alpha}^{\varphi'}$ and possesses an index j such that $\varphi_j'(x) \le h\alpha(x)$ almost everywhere. However, no confusion should arise because of our more informal wording.) Hence, f_j also belongs to $C_\alpha^{\varphi'}$, or $\varphi_j'(x) \le \alpha(x)$ almost everywhere. By (15), $\varphi_j(x) \le h\alpha(x)$ almost everywhere, which implies that f_j belongs to $C_{h\alpha}^{\varphi}$. Hence, we have shown that

$$C_{h\alpha}^{\varphi'} \subseteq C_{h\alpha}^{\varphi}.$$

This result, although surprising, does not necessarily have practical implications, the reason being the condition *almost everywhere* in the definition of complexity classes.

The gap theorem should be contrasted with the subsequent **compression theorem.** Consider complexity classes C_α, where the recursive function α is also a *cost* function. According to the compression theorem, we can construct a recursive function α_1 uniform effectively in terms of α such that $C_\alpha \subsetneq C_{\alpha_1}$. Moreover, there are functions assuming small values and belonging to the difference $C_{\alpha_1} - C_\alpha$. Consequently, the *optimal* algorithms for such functions become computable if the function limiting the resource is increased from α to α_1.

Theorem 6.9. *There is a recursive function e of two variables such that, for every recursive cost function α,*

$$C_\alpha \subsetneq C_{\alpha_1} \quad \text{where} \quad \alpha_1(x) = e(x, \alpha(x)).$$

Furthermore, there is a recursive function h such that the conditions (i) and (ii) are satisfied for every total recursive cost function $\alpha = \varphi_i$.

 (i) *Whenever* $f_j = f_{h(i)}$, *then* $\varphi_j(x) > \varphi_i(x)$ *almost everywhere.*
 (ii) $\varphi_{h(i)}(x) \leq e(x, \varphi_i(x))$ *almost everywhere.*

Consequently, the function $f_{h(i)}$ is in the difference $C_{\alpha_1} - C_\alpha$.

Proof.   Consider the function $\psi(x, i)$ defined by

$$
\psi(x, i) = \begin{cases}
\max \{f_v(x) \mid v \leq x \text{ and } \varphi_v(x) \leq \varphi_i(x)\} + 1 \\
\quad \text{if } \varphi_i(x)\!\downarrow \text{ and some } v \text{ satisfying the conditions exists,} \\
0 \quad \text{if } \varphi_i(x)\!\downarrow \text{ and no } v \text{ satisfying the conditions exists,} \\
\uparrow \quad \text{if } \varphi_i(x)\!\uparrow.
\end{cases}
$$

Clearly, the function $\psi(x, i)$ is partial recursive: We first test step by step whether $\varphi_i(x)\!\downarrow$, and if a positive answer is found, the condition $\varphi_v(x) \leq \varphi_i(x)$ can be tested by Axiom B2 for all $v \leq x$. By our standard *s-m-n* construction, a recursive function h satisfying $f_{h(i)}(x) = \psi(x, i)$ for all i is found. Moreover, if φ_i is total, then $f_{h(i)}$ will be total. From now on we assume that φ_i is total. (In fact, this is the only alternative considered in the statement of the theorem.)

Before defining the function e, we claim that the function h satisfies condition (i). Assume the contrary: For some j satisfying $f_j = f_{h(i)}$, there are infinitely many values of x such that $\varphi_j(x) \leq \varphi_i(x)$. Hence, $\varphi_j(x_1) \leq \varphi_i(x_1)$ holds for some $x_1 \geq j$. On the other hand, by the definition of ψ,

$$f_{h(i)}(x_1) = \max \{f_v(x_1) \mid v \leq x_1 \text{ and } \varphi_v(x_1) \leq \varphi_i(x_1)\} + 1.$$

Consequently, because $x_1 \geq j$, we obtain

$$f_j(x_1) = f_{h(i)}(x_1) \geq f_j(x_1) + 1,$$

which is a contradiction establishing our claim that h satisfies condition (i). This shows also that $f_{h(i)}$ does not belong to C_α for $\alpha = \varphi_i$.

Now define the function e_1 by

$$
e_1(i, x, y) = \begin{cases}
\varphi_{h(i)}(x) & \text{if } \varphi_i(x) = y \\
0 & \text{otherwise.}
\end{cases}
$$

Clearly, e_1 is recursive. (Here the assumption that φ_i is total is not needed.) Next, define the function e by

$$e(x, y) = \max_{i \leq x} e_1(i, x, y).$$

Hence, if $\varphi_i(x)\!\downarrow$ and $x \geq i$, we obtain

$$\varphi_{h(i)}(x) = e_1(i, x, \varphi_i(x)) \leq e(x, \varphi_i(x)).$$

This shows that condition (ii) is satisfied. Moreover, it shows that $f_{h(i)}$ is in C_{α_1}, which in turn implies the relation

$$C_\alpha \subsetneqq C_{\alpha_1}. \qquad \square$$

By a slightly more sophisticated diagonalization technique, a further condition to the effect that the functions $f_{h(i)}$ assume only small values can be satisfied (Exercise 6.4.).

6.3. SPEEDUP THEOREM: FUNCTIONS WITHOUT BEST ALGORITHMS

A rather surprising result is established in this section. No matter what complexity measure and amount of speedup we consider, there are recursive functions f such that an arbitrary algorithm for f can be sped up by that amount. Suppose $f = f_i$ and we consider the algorithm determined by f_i (for instance, the Turing machine TM_i) to be a particularly efficient one. This means that we consider the amount of resource defined by φ_i to be particularly small, in view of the complexity of the function f. Suppose, further, that $h(x)$ is a very rapidly increasing function, for instance,

$$h(x) = 2^{2^{2^{2^x}}}.$$

Then there is another algorithm for f using so much less resource as h indicates. In other words, we have also $f = f_j$ and

$$\varphi_i(x) \geq 2^{2^{2^{2^{\varphi_j(x)}}}}. \tag{16}$$

However, (16) does not necessarily hold for all x but only almost everywhere. Of course, the same procedure can be repeated for f_j. This gives rise to another speedup analogous to (16), but now f_j is the algorithm using "much" resource. The speedup can be repeated again for the resulting function, and so forth ad infinitum.

Thus, the speedup theorem tells us that some functions f have no best algorithms: For every person's favorite algorithm f_i, a speedup such as (16) is possible. However, this is only true for some functions f that might be considered "unnatural." At the end of this section, we return to the problem of when a speedup is possible. But first we establish the following **speedup theorem.**

Theorem 6.10. *For every recursive function $g(x, y)$ (no matter how rapidly growing), there is a recursive function $f(x)$ such that for every index i of f, there is an index j of f with the property that*

$$g(x, \varphi_j(x)) < \varphi_i(x) \tag{17}$$

holds almost everywhere.

Proof. We use the notation

$$x \dotminus i = \begin{cases} x - i & \text{if } x - i \geq 0 \\ 0 & \text{if } x - i < 0. \end{cases}$$

We shall show the existence of recursive functions $f(x)$ and $h(x)$ such that for every index i of f, there is an index j of f with the property that the conditions

$$\varphi_i(x) > h(x \div i) \tag{18}$$

and

$$g(x, \varphi_j(x)) \leq h(x \div i) \tag{19}$$

hold almost everywhere. Clearly, (17) is a consequence of (18) and (19).

The main part of the proof of Theorem 6.10 consists of establishing the following lemma. In the statement of the lemma, we refer to the *s-m-n* theorem. Observe that the *s-m-n* theorem applies for any variables of the function instead of the last ones, to which we usually have applied it.

Lemma 6.11.   *There is a partial recursive function $C(u, v, t, x)$ such that conditions (i)–(iii) are satisfied.*

(i) *Whenever f_t is recursive, so is the function obtained by the s-m-n theorem*

$$f_{s(u,v,t)}(x) = C(u, v, t, x) \tag{20}$$

for all u and v and, moreover, for every u, there is a v_u such that

$$C(u, v_u, t, x) = C(0, 0, t, x)$$

holds for every x.

(ii) *Assume that f_t is recursive and define*

$$f(x) = f_{s(0,0,t)}(x) = C(0, 0, t, x). \tag{21}$$

Then, for every index i of f,

$$\varphi_i(x) > f_t(x \div i)$$

holds almost everywhere.

(iii) *There is a recursive function $h(x) = f_t(x)$ such that if f is defined by (21) and i is an arbitrary index of f and we choose $u = i + 1$, $v = v_{i+1}$, then for $j = s(u, v, t)$,*

$$h(x \div i) \geq g(x, \varphi_j(x)) \tag{22}$$

holds almost everywhere in x.

We show first that Theorem 6.10 is a consequence of Lemma 6.11. Thus, assume that a function C satisfying (i)–(iii) exists. Consider the function $h = f_t$. Let f now be defined by (21), and let i be an arbitrary index of f. Then (ii) guarantees that (18) is satisfied. On the other hand, (iii) guarantees that (19) is satisfied because (21) and (i) imply that j is also an index of f. Hence, Theorem 6.10 follows.

We now begin the proof of Lemma 6.11, first giving the definition of the function $C(u, v, t, x)$.

Case 1. Assume that either $u = 0$ or $x \geq v$. If i is an index satisfying condition A stated below, we define

$$C(u, v, t, x) = f_i(x) + 1$$

and *cancel* the index i. Whenever we have established that no index i satisfies condition A, we define

$$C(u, v, t, x) = 0.$$

Case 2. Assume that both $u > 0$ and $x < v$. Then we define

$$C(u, v, t, x) = C(0, 0, t, x).$$

Condition A. i is the smallest index not canceled when defining the values

$$C(u, v, t, y) \quad \text{with} \quad v \leq y < x, \quad \text{or} \quad C(0, 0, t, y) \quad \text{with} \quad 0 \leq y < v, \quad (23)$$

and satisfying

$$u \leq i \leq x \quad \text{and} \quad \varphi_i(x) \leq f_i(x \div i). \quad (24)$$

(It is understood that both sides are defined in order for the inequality to hold.)

Clearly, the function C so defined is partial recursive. Observe that when computing the value $C(u, v, t, x)$, we must first compute the values in (23). (If $v = 0$, the set of the latter values in (23) is empty.) Of course, C may be divergent for some argument values because we might never find out the validity of condition A.

We still have to show that C satisfies conditions (i)–(iii) of Lemma 6.11. The reader might wonder at this point how it is possible that C does not depend at all on g. The reason is that the function h in (iii) depends on g, and the definition of C provides enough leeway to construct such an h independently of g.

To establish (i), assume that f_i is recursive. We can then check condition A effectively and, consequently, function (20) is recursive for all u and v. (This is also true if it is defined via case 2.) Consider a fixed value of u. We have to show the existence of a number v_u such that

$$C(u, v_u, t, x) = C(0, 0, t, x) \quad (25)$$

holds for every x.

If $u = 0$, we may choose $v_u = 0$. Hence, assume that $u > 0$. Let v_u be an integer such that if an index $i \leq u$ is canceled when defining some of the values $C(0, 0, t, x)$, it is canceled when defining some of the values $C(0, 0, t, x)$ with $x < v_u$. Such an integer v_u exists, since every index is canceled at most once in the course of the definition of the values $C(0, 0, t, x)$. We claim that (25) holds for every x.

Assume the contrary, and let $x = x_1$ be some value not satisfying (25). By case 2 in the definition of C, we must have $x_1 \geq v_u$, since $u > 0$. Suppose first that $C(0, 0, t, x_1) = 0$—that is, there is no index i satisfying condition A for the argument $(0, 0, t, x_1)$. However, it is easy to see that this implies that there is no index i satisfying condition A for the argument (u, v_u, t, x_1). Hence, $C(u, v_u, t, x_1) = 0$, which contradicts the choice of x_1.

Suppose, secondly, that $C(0, 0, t, x_1) > 0$. Let k be the index satisfying condition A for the argument $(0, 0, t, x_1)$. We must have $k > u$ because all indices less than or equal to u have been canceled when defining the values $C(0, 0, t, x)$ with $x < v_u$ (provided they are canceled when defining some value $C(0, 0, t, x)$).

We claim that k also satisfies condition A for the argument (u, v_u, t, x_1), which again contradicts the choice of x_1. We already know that (24) is satisfied for $i = k$ and $x = x_1$. We also know that k is not canceled when defining the values $C(0, 0, t, y)$ with $y < v_u$. If it were canceled when defining some value $C(u, v_u, t, y)$ with $v_u \leq y < x_1$, it would also have to be canceled when defining some value $C(0, 0, t, y')$ with $y' < x_1$, which is impossible. Finally, if k were not smallest for the argument (u, v_u, t, x_1), it would not be smallest for the argument $(0, 0, t, x_1)$ either. Hence, we have shown that C satisfies condition (i) of Lemma 6.11.

The proof of condition (ii) is easy. Consider an arbitrary index i of f. If the index i were canceled when defining some value $C(0, 0, t, x)$, Case 1 of the definition of C would give the impossible equation $f(x) = f_i(x) + 1$. Hence, i is never canceled when defining the values $C(0, 0, t, x)$. Consider indices $i' < i$ that are canceled when defining the values $C(0, 0, t, x)$, and let x' be such that all these indices are canceled when defining the values $C(0, 0, t, x)$ with $x \leq x'$. Then

$$\varphi_i(x) > f_t(x \div i)$$

holds for $x > x'$.

To establish condition (iii), first consider the function H of two variables defined by

$$H(0, k) = 0,$$

$$H(x + 1, k) = \begin{cases} \max \{g(x + u, \varphi_{s(u,v,k)}(x + u)) \mid 0 \leq u,\, v \leq x\} \\ \quad \text{if } f_k(y){\downarrow} \text{ for every } y \leq x \\ \uparrow \quad \text{otherwise.} \end{cases}$$

Since H clearly is partial recursive, the s-m-n theorem (Theorem 4.1) guarantees the existence of a recursive function $\bar{s}$ such that $H(x, k) = f_{\bar{s}(k)}(x)$. On the other hand, the recursion theorem (Theorem 4.6) guarantees the existence of a number t such that $f_{\bar{s}(t)} = f_t$. Hence, by the definition of H,

$$f_t(0) = 0,$$

$$f_t(x + 1) = \begin{cases} \max\ \{g(x + u, \varphi_{s(u, v, t)}(x + u)) \mid 0 \le u, v \le x\} \\ \qquad \text{if } f_t(y)\!\downarrow \text{ for every } y \le x \\ \uparrow \quad \text{otherwise.} \end{cases}$$

We shall prove that f_t is total, which implies that it is recursive. We show by induction on x that $f_t(x)\!\downarrow$, for every x. Clearly, $f_t(0)\!\downarrow$. Assume that $f_t(y)\!\downarrow$ for every $y \le x$. Then, by the above considerations, we have $f_t(x + 1)\!\downarrow$ if

$$C(u, v, t, x + u)\!\downarrow \quad \text{for} \quad 0 \le u, v \le x. \tag{26}$$

But (26) holds because condition A can be tested for the arguments involved. Indeed, for these arguments, only values $f_t(y)$ with $y \le x$ appear in (24). But for these values, convergence has already been established, which completes the induction.

We now let $f_t = h$. Let f be defined by (21) and let i be an arbitrary index of f. Assume that

$$x > \max\ \{v_{i+1} + (i + 1), 2(i + 1)\}. \tag{27}$$

Then

$$h(x \dotminus i) = f_t(x \dotminus i)$$
$$= \max\{g(x + u \dotminus (i + 1), \varphi_{s(u,v,t)}(x + u \dotminus (i + 1))) \mid 0 \le u, v \le x \dotminus (i + 1)\}.$$

By substituting $u = i + 1$, $v = v_{i+1}$, we obtain

$$h(x \dotminus i) \ge g(x, \varphi_{s(i+1, v_{i+1}, t)}(x)). \tag{28}$$

(Observe that the u and v chosen lie in the proper ranges because of (27).) Since (28) holds for all values of x satisfying (27), we conclude that (22) is true almost everywhere in x for $j = s(i + 1, v_{i+1}, t)$. This completes the proof of (iii) and, at the same time, the proof of Lemma 6.11 and Theorem 6.10. $\square$

Although the proof of Theorem 6.10 is quite complicated, the key ideas are simple. The definition of C guarantees that (18) holds almost everywhere in x, no matter which index i of f we choose. Then, essentially, we find a particular index j such that (19) holds with i replaced by $i + 1$.

The speedup theorem has the following interpretation in terms of the computers FOSSIL and FUTURA discussed earlier. There are functions f such that for every program f_i for f, there is another program f_j for f such that f_j computes f faster (almost everywhere) on FOSSIL than f_i computes f on FUTURA. Indeed, if φ_i and φ_i' again are the cost functions determined by the computing times of FOSSIL and FUTURA, respectively, then there is a recursive function g such that

$$\varphi_i(x) \leq g(x, \varphi_i'(x))$$

holds for all i and x. Now apply the speedup theorem for this g. Hence, there are functions f such that for every index i of f, there is an index j of f with the property that

$$\varphi_j(x) \leq g(x, \varphi_j'(x)) < \varphi_i'(x)$$

holds almost everywhere in x.

The proof of the speedup theorem is not effective. For instance, the construction of the numbers v_u is not effective (see (i) in Lemma 6.11). In fact, every proof of the speedup theorem must be noneffective. This is a consequence of the following result, mentioned here without further discussion.

Let g be a sufficiently fast-growing recursive function, and let f be an arbitrary speedable function (whose existence is guaranteed by Theorem 6.10). Then there is no partial recursive function r such that whenever i is an index for f, then $r(i)$ converges and satisfies

$$g(x, \varphi_{r(i)}(x)) < \varphi_i(x)$$

almost everywhere in x.

The condition *almost everywhere* cannot be removed from the speedup theorem. Indeed, it can be shown that there is no recursive bound for the exceptional values of the argument x. We can also study the complexity of an algorithm in terms of the number of instructions in a program executing the algorithm. Then we can show that there is no recursive bound for the complexity (in this static sense) of the faster algorithm.

We shall next establish a version of the speedup theorem corresponding to the representation developed in connection with Hilbert's tenth problem.

Consider a polynomial $P(x, y_1, \ldots, y_m)$ with integer coefficients. Denote by $Y(x)$ the set of m-tuples $(y_1, \ldots, y_m)$ of nonnegative integers such that

$$P(x, y_1, \ldots, y_m) = 0. \tag{29}$$

Define

$$\varphi(x, P) = \min_{(y_1, \ldots, y_m) \in Y(x)} \max\{y_1, \ldots, y_m\}.$$

Of course, $Y(x)$ may be empty, in which case $\varphi(x, P)$ is considered to be undefined.

Let $P_1, P_2, \ldots$ be an effective enumeration of all polynomials with integer coefficients and with at least three variables. (Thus, the indexing P_i should be such that we can go effectively from a polynomial P to its index, and vice versa.) Then m-tuples $(y_1, \ldots, y_m)$ are ordered using the pairing function φ_m presented in Section 4.3:

$$(y_1, \ldots, y_m) < (y_1', \ldots, y_m') \quad \text{iff} \quad \varphi_m(y_1, \ldots, y_m) < \varphi_m(y_1', \ldots, y_m'). \tag{30}$$

A polynomial $P_i = P(x, y_1, \ldots, y_m)$ defines a partial recursive function $f_i(x)$ as follows. The value $f_i(x)$ equals the number y_1 in the smallest (in the sense of (30)) m-tuple in $Y(x)$. If $Y(x)$ is empty, then $f_i(x)$ is undefined. Clearly, the functions $f_i(x)$ defined in this fashion are partial recursive. Moreover, it is easy to see that the enumeration $f_1, f_2, \ldots$ thus defined constitutes an acceptable enumeration of all partial recursive functions (Exercise 6.7).

We now define $\varphi_i(x) = \varphi(x, P_i)$. The sequences f_i and φ_i constitute a complexity measure. Indeed, it is clear that Axiom B1 is satisfied. Moreover, the validity of the equation $\varphi_i(x) = z$ can be determined by checking the validity of (29) for finitely many m-tuples $(y_1, \ldots, y_m)$. Hence, Axiom B2 is also satisfied.

A pair of polynomials (P_1, P_2) is referred to as a **characteristic pair** for a recursive set S iff

$$S = \{x \mid (\exists y_1 \geq 0) \ldots (\exists y_m \geq 0)(P_1(x, y_1, \ldots, y_m) = 0)\}$$

and

$$\sim S = \{x \mid (\exists y_1 \geq 0) \ldots (\exists y_m \geq 0)(P_2(x, y_1, \ldots, y_m) = 0)\}.$$

(It is no loss of generality to assume, as is done in this definition, that the polynomials in a characteristic pair have the same number of variables.)

Theorem 6.12. *Let $g(x, y)$ be a recursive function. Then there is a recursive set S such that for every characteristic pair (P_1, P_2) for S, there is another characteristic pair (Q_1, Q_2) for S with the property that almost everywhere in x*

$$x \in S \quad implies \quad \varphi(x, P_1) > g(x, \varphi(x, Q_1))$$

and

$$x \notin S \quad implies \quad \varphi(x, P_2) > g(x, \varphi(x, Q_2)).$$

Proof. Consider the complexity measure determined by the sequences f_i and φ_i discussed before Theorem 6.12. Let f be a function assuming only the values 0 and 1 (see Exercise 6.6), obtained by applying the speedup theorem for the given function g. Define

$$S = \{x \mid f(x) = 0\}. \tag{31}$$

Since f is recursive, we obtain

$$\sim S = \{x \mid f(x) = 1\}. \tag{32}$$

Let (P_1, P_2) be an arbitrary characteristic pair for S. (Indeed, Theorem 5.9 guarantees the existence of a characteristic pair for every recursive set.) Consider the polynomial

$$P = [(1 - y)P_1 + yP_2]^2 + y^2(y - 1)^2,$$

where it is understood that y is the first variable after the variable x. Compute the index of P: Assume that $P = P_i$.

We claim that $f = f_i$. Assume that $x, y, y_1, \ldots, y_m$ are the variables of P. Since (P_1, P_2) is a characteristic pair, there is no assignment for the variables $x, y_1, \ldots, y_m$ satisfying

$$P_1(x, y_1, \ldots, y_m) = P_2(x, y_1, \ldots, y_m) = 0.$$

Consequently, $P = 0$ for those assignments only for which either $y = 1$ and $P_2 = 0$ or $y = 0$ and $P_1 = 0$. Moreover, the value of y equals the value of $f_i(x)$, where x comes from the corresponding $(m + 1)$-tuple. Observe that the second term of P guarantees that P can equal 0 only if $y = 0$ or $y = 1$. Hence, $f = f_i$.

By the speedup theorem, there is a polynomial $Q(x, y, y_1, \ldots, y_n)$ such that if j is the index of Q, then $f = f_j$ and

$$\varphi_i(x) > g(x, \varphi_j(x)) \tag{33}$$

holds almost everywhere. Now define

$$Q_1(x, y_1, \ldots, y_n) = Q(x, 0, y_1, \ldots, y_n)$$

and

$$Q_2(x, y_1, \ldots, y_n) = Q(x, 1, y_1, \ldots, y_n).$$

If x is in S, then $\varphi(x, P_1)\downarrow$ and $\varphi(x, Q_1)\downarrow$. Moreover, by (33),

$$\varphi(x, P_1) > g(x, \varphi(x, Q_1))$$

holds almost everywhere. Similarly, if x is not in S, then $\varphi(x, P_2)\downarrow$ and $\varphi(x, Q_2)\downarrow$ and

$$\varphi(x, P_2) > g(x, \varphi(x, Q_2))$$

holds almost everywhere.   □

The function f of the speedup theorem has no best algorithm because every algorithm for f can be sped up the amount indicated by the given recursive function g. It is clear that such a speedup is not possible for every function, even if it happens under the condition *almost everywhere*. There certainly are functions with best algorithms and, thus, with no speedup —typical examples being constant functions. In this sense, we may speak of *speedable* and *nonspeedable* functions. We now present, without proofs, some results from this area. It will be more convenient to express the notions in terms of recursively enumerable sets rather than recursive functions. This essentially extends the notion of speedup from total to partial functions. It is understood that a specific complexity measure underlies the definitions.

A recursively enumerable set S (of nonnegative integers) is **speedable** iff for all indices i such that $S = S_i$ and for all recursive functions $g(x, y)$, there exists an index j such that $S = S_j$ and

$$\varphi_i(x) > g(x, \varphi_j(x))$$

holds for infinitely many values of x in S. Furthermore, S is **effectively speedable** iff j can be effectively computed from i and an index for g.

Let us define the **semicharacteristic** function for a set S to be the partial function $f(x)$ assuming the value 1 for x in S and being undefined for x not in S. Then, intuitively, a set S is effectively speedable if from any program for the semicharacteristic function of S, we can effectively obtain another program for the same function that is arbitrarily faster than the original program on infinitely many integers. If a set is speedable, then only the existence of the faster program is known. We can show that there are sets which are speedable but not effectively speedable.

The notion complementary to speedability is now defined: A recursively enumerable set S is **nonspeedable** iff there is an index k for S and a recursive function $g(x, y)$ such that whenever j is an index for S, the condition

$$x \in S \quad \text{implies} \quad \varphi_k(x) \le g(x, \varphi_j(x))$$

holds almost everywhere in x.

The following theorem, whose proof is omitted, characterizes nonspeedable sets.

Theorem 6.13. *A recursively enumerable set S is nonspeedable iff there is a recursive function $f(x)$ such that for all indices i, $S_i \cap \sim S = S_{f(i)} \cap \sim S$ and, whenever $S_i \subseteq S$, then $S_{f(i)}$ is finite.*

A characterization of effective speedability can be given in terms of subcreative sets. By definition, a recursively enumerable set S is **subcreative** iff there is a recursive function $f(x)$ with the property that for every index i such that S_i and S are disjoint, we have $S \subsetneq S_{f(i)} \subseteq \sim S_i$. It is easy to see that no recursive set is subcreative and that every creative set is subcreative. There are subcreative sets that are not creative, and all subcreative sets are Turing complete. The proofs of these assertions, as well as the proof of the following theorem, are omitted.

Theorem 6.14. *A recursively enumerable set is effectively speedable iff it is subcreative.*

The definition above also introduces the notion of effective speedability for partial recursive functions in a natural fashion. Theorem 6.14 then tells us that a partial recursive function $f(x)$ is effectively speedable iff the set of pairs $(x, f(x))$ is subcreative. (We may, of course, use the pairing function to encode such pairs as numbers.) This shows, for instance, that cost functions can never be effectively speedable.

Notions of speedability are **measure-independent:** If a set is speedable (resp. effectively speedable, nonspeedable) in one measure, it is so in all measures. We now exhibit a complexity-theoretic property that is not measure-independent.

If a recursively enumerable set S fails to have a *single* fastest program, it might still be possible to find a computable *sequence* of partial recursive functions that are cofinal in the cost evaluation of all programs for S. This leads to the following definition.

A recursively enumerable set S has a **Meyer-Fischer complexity sequence** iff there is a recursive function $g(x)$ such that conditions (i)–(iii) hold for every index i.

(i) $S_{g(i)} = S$.

(ii) Whenever $S_i = S$, there is a j such that

$$f_{g(j)}(x) \le \varphi_i(x) \qquad \text{almost everywhere.}$$

(iii) There is a k such that $S_k = S$ and

$$\varphi_k(x) \le f_{g(i)}(x) \qquad \text{almost everywhere.}$$

(In fact, condition (i) is unnecessary because it is a consequence of (ii) and (iii). We have added it for clarity.)

We can show that the property of possessing a Meyer-Fischer complexity sequence is measure-dependent. Once we know that there are both measure-independent and measure-dependent properties, it is fairly easy to show that there is no algorithm for deciding the measure-independence of complexity-theoretic properties.

6.4. TIME BOUNDS, THE CLASSES $\mathcal{P}$ AND $\mathcal{NP}$, AND $\mathcal{NP}$-COMPLETE PROBLEMS

In the remainder of this chapter we study specific complexity measures, defined in terms of computing time and memory space. Although the abstract theory based on Axioms B1 and B2 is, of course, also valid in this case, we can use terms of the specific measures to express more-detailed ideas concerning notions, such as intractability. Also, the problem of whether or not $\mathcal{P}$ equals $\mathcal{NP}$, which is perhaps the most celebrated open problem in the theory of computation, belongs to this area.

In the abstract complexity theory, a cost function φ_i is associated with a partial recursive function f_i. The value $\varphi_i(n)$ is interpreted as the cost, or the amount of resource, for computing $f_i(n)$. Thus, the amount of resource depends on the size of the input. This is the approach we follow also now. For instance, by a **polynomial time bound** for an algorithm we mean a polynomial $P(n)$ such that the algorithm uses at most $P(n)$ units of time when operating on an input of size n.

Most of the problems we consider will be decision problems, having a yes or no answer. In other words, the functions computed can be viewed as characteristic functions for some sets. In this way, we avoid awkward situations where the answer to our question (that is, the value of the function to be computed if the setup is our usual one) occupies an enormous amount of space in comparison with the size of the input. In such situations the writing of the output could take an amount of time and space exponential in terms of the length of the input. Hence, it is a priori clear that there cannot exist any algorithm with a polynomial time bound for such a problem. The problem would be a priori intractable according to the definitions given below. Before the definitions, we discuss a couple of examples. They provide some background for the notions to be defined and are also, as we shall see, basic for the theory of $\mathfrak{N}\mathcal{P}$-completeness.

Example 6.1. Consider the alphabet

$$\Sigma = \{1, 2, \vee, \wedge, \sim, (,)\}.$$

A word w over the alphabet Σ is a **well-formed formula of the propositional calculus**, abbreviated **wffpc**, iff either (i) or (ii) holds:

(i) w is a nonempty word over the alphabet $\{1, 2\}$.
(ii) There are wffpc's u and v such that

$$w = (u \vee v), \text{ or } w = (u \wedge v), \text{ or } w = \sim u.$$

Intuitively, $\vee$, $\wedge$, and $\sim$ stand for disjunction, conjunction, and negation. We may have in wffpc's unboundedly many variables x_i, where i is an integer in dyadic notation. Thus, instead of x_9 we write 121. Condition (i) says that every variable alone is a wffpc.

Formally, every subword $\alpha \in \{1, 2\}^+$ of a wffpc w satisfying

$$w = w_1 \alpha w_2, \quad w_1 \notin \Sigma^*\{1, 2\}, \quad w_2 \notin \{1, 2\}\Sigma^*$$

is called a **variable**.

Let $\alpha_1, \ldots, \alpha_n$ be all the variables appearing in a wffpc w. A mapping T of the set $\{\alpha_1, \ldots, \alpha_n\}$ into the set $\{0, 1\}$ is termed a **truth-value assignment** for w. The **truth value** of w for the assignment T is determined according to the following rules. The truth value of a variable α_i equals $T(\alpha_i)$. The truth value of $(u \vee v)$ (resp. $u \wedge v$) equals $\max(u_1, v_1)$ (resp. $\min(u_1, v_1)$), where u_1 and v_1 are the truth values of u and v, respectively. The truth value of $\sim u$ equals $1 - u_1$, where u_1 is the truth value of u.

A wffpc w is **satisfiable** iff it assumes the truth value 1 for some truth-value assignment T. We denote the language over Σ consisting of all satisfiable wffpc's by SAT.

Intuitively, 1 and 0 stand for the truth values *true* and *false,* respectively. A wffpc is satisfiable iff it is not identically false under the familiar truth-table technique. Of course, in every truth-value assignment, all occurrences of a particular variable α_i get the same truth-value.

In what follows, the strict rules of the definition of a wffpc are somewhat relaxed. We use lowercase letters from the end of the alphabet to denote variables. Thus, a particular variable might be denoted by x_9 instead of the notation 121 specified in the definition. Unnecessary parentheses are omitted. This applies also for parentheses unnecessary because of the associativity of $\vee$ and $\wedge$. (We are interested only in truth values and, clearly, the functions min and max are associative.)

Consider the following two wffpc's:

$$(x_1 \vee \sim x_2) \wedge (\sim x_1 \vee x_2) \wedge (\sim x_1 \vee \sim x_2) \wedge x_3, \tag{34}$$

$$(x_1 \vee \sim x_2 \vee x_3) \wedge (x_2 \vee x_3) \wedge (\sim x_1 \vee x_3) \wedge \sim x_3. \tag{35}$$

Both (34) and (35) are conjunctions of wffpc's, each of which is a disjunction of **literals,** where variables and their negations are referred to as literals. We say that wffpc's of this type are in **conjunctive normal form**. Moreover, if every disjunction contains at most three (resp. two) literals, we say that the wffpc is in **3-conjunctive** (resp. **2-conjunctive**) normal form. Thus, (35) is in 3-conjunctive normal form, and (34) is in 2-conjunctive (as well as in 3-conjunctive) normal form.

We denote by CONSAT the language over Σ consisting of all satisfiable wffpc's in conjunctive normal form. The notations 3-CONSAT and 2-CONSAT are defined analogously. The wffpc (34) is in 2-CONSAT, but the wffpc (35) is not in 3-CONSAT because it is not at all satisfiable. This is seen by the following argument. The last clause in (35) forces the assignment 0 for x_3. Hence, the second and third clauses force the assignments 1 and 0 for x_2 and x_1, respectively. But now the first clause assumes the value 0 for this assignment.

Clearly, satisfiability is a decidable property. We just have to check through all of the 2^n truth-value assignments possible for n variables. (This is, in fact, what the familiar truth-table technique amounts to.) Such an exhaustive check uses an exponential amount of time (in terms of the number of variables or the length of the given wffpc). We now describe briefly an algorithm for testing satisfiability that is based on the reduction of the number of variables. It is assumed that the input is given in conjunctive normal form. This algorithm exhibits a remarkable difference between 2-conjunctive and 3-conjunctive normal forms. The difference will be discussed later in connection with time complexity.

Assume that α is a wffpc in conjunctive normal form. Thus,

$$\alpha = \alpha_1 \wedge \alpha_2 \wedge \cdots \wedge \alpha_k,$$

where each α_i is a disjunction of literals. We refer to the disjunctions α_i as **clauses.**

Step 1. Make sure that every variable occurs (either negated or unnegated) at most once in each clause. This is accomplished by making the following changes in α. Every clause containing both x and $\sim x$, for some variable x, is removed from α. If x (resp. $\sim x$) occurs several times in some clause, these occurrences are replaced by a single occurrence of x (resp. $\sim x$). If everything is removed, α is satisfiable. (In fact, it is identically true.) Otherwise, let α' be the resulting wffpc.

Step 2. Replace α' by a wffpc α'' containing no clauses with only one literal (and also satisfying the condition required from α' after Step 1). Indeed, if x (resp. $\sim x$) occurs alone in some clause, remove all clauses containing x (resp. $\sim x$) and, furthermore, remove $\sim x$ (resp. x) from every clause where it occurs together with some other variable; if $\sim x$ (resp. x) occurs alone in some other clause, conclude that α is not satisfiable. Iterate the procedure until α'', described above, is obtained.

Step 3. If no variable occurs in α'' both negated and unnegated, conclude that α is satisfiable. Otherwise, choose a variable x such that both x and $\sim x$ occur in α''. Find all the clauses

$$(x \vee \beta_1), \ldots, (x \vee \beta_m), (\sim x \vee \gamma_1), \ldots, (\sim x \vee \gamma_n)$$

where x or $\sim x$ occurs. Let δ be the conjunction of the other clauses (if any). Then α'' is satisfiable iff the wffpc

$$((\beta_1 \wedge \cdots \wedge \beta_m) \vee (\gamma_1 \wedge \cdots \wedge \gamma_n)) \wedge \delta$$

is satisfiable. Observe that each of the β's and γ's contains at least one literal and contains exactly one literal if the original α is in 2-conjunctive normal form.

If some of the β's or γ's contain more than one literal, replace α'' by the two wffpc's

$$\bar{\alpha} = \beta_1 \wedge \cdots \wedge \beta_m \wedge \delta \quad \text{and} \quad \bar{\bar{\alpha}} = \gamma_1 \wedge \cdots \wedge \gamma_n \wedge \delta,$$

make sure that neither $\bar{\alpha}$ nor $\bar{\bar{\alpha}}$ contains the same clause twice (remove unnecessary occurrences), and return to Step 1. The original α is satisfiable iff $\bar{\alpha}$ or $\bar{\bar{\alpha}}$ is satisfiable.

If every β and γ contains exactly one literal, replace α'' by the wffpc

$$\alpha''' = (\beta_1 \vee \gamma_1) \wedge \cdots \wedge (\beta_1 \vee \gamma_n) \wedge \cdots \wedge (\beta_m \vee \gamma_n) \wedge \delta,$$

remove duplicate occurrences of the same clause, and return to Step 1. The original α is satisfiable if α''' is.

This concludes the description of the algorithm. The reader should have no difficulties in verifying that the method works. Some explanations were already given above. The essential point is that the number of variables is properly decreased before each return to Step 1.   ☐

The discussion in Example 6.1 is quite long because the languages CONSAT and 3-CONSAT are of crucial importance later.

Example 6.2. Consider words of the form

$$w_0 \# w_1 \# \cdots \# w_k$$

over the alphabet $\{1, 2, \#\}$ such that $k \geq 1$, each of the w's is a nonempty word over the alphabet $\{1, 2\}$, and, moreover, w_0 equals the sum of some of the other w's when the words are viewed as dyadic integers. We denote by KNAPSACK the language consisting of all such words. $\square$

We could have denoted the variables in Example 6.1 by unary notation $(1, 11, 111, \ldots)$ without affecting our forthcoming discussions about $\mathfrak{N}\mathcal{P}$-completeness. However, in Example 6.2 it is quite essential that the notation is dyadic: KNAPSACK as defined is $\mathfrak{N}\mathcal{P}$-complete, whereas the unary version of the same language is not $\mathfrak{N}\mathcal{P}$-complete.

We are now ready for the basic definitions. Consider a Turing machine TM that halts with all inputs w. The **time-complexity function** associated with TM is defined by

$$t_{TM}(n) = \max\{m \mid TM \text{ halts after } m \text{ steps with an input } w \text{ such that } |w| = n\}.$$

If TM is defined in terms of a rewriting system, as done in Chapter 4, then the number of steps equals the length of the derivation according to the rewriting system. However, in this chapter any other customary formalization of Turing machines may also be considered. Of course, if the rewriting model of Chapter 4 is considered, the mode of acceptance has to be modified: We do not want all words to be accepted!

Thus $t_{TM}(n)$ maps the set of nonnegative integers into itself. We say that TM is **polynomially bounded** iff there is a polynomial $P(n)$ such that $t_{TM}(n) \leq P(n)$ holds for all n. We denote by $\mathcal{P}$ the family of languages acceptable by polynomially bounded Turing machines. (So far we have considered only deterministic Turing machines—nondeterministic ones are introduced later.)

Although $\mathcal{P}$ is defined as a family of languages, it can be visualized as the collection of problems for which there exists an algorithm operating in polynomial time. This observation has already been made before: We can identify a decision problem with the membership problem for the language of "positive instances." Thus, the satisfiability problem of wffpc's discussed in Example 6.1 is (at least essentially) the same as the membership problem for the language SAT. To decide whether or not "a knapsack of size w_0 can be filled with some of the items of sizes $w_1, \ldots, w_k$" amounts to deciding membership in the language KNAPSACK defined in Example 6.2.

The family $\mathcal{P}$ is very natural from a mathematical point of view. This is seen from the fact that it is highly invariant with respect to the underlying model of computation. For instance, Turing machines TM_1 with several tapes are faster than ordinary Turing machines—that is, their time-complexity function assumes smaller values. However, if the time-complexity function of such a TM_1 is bounded from above by a polynomial $P_1(n)$, one can (effectively) construct an ordinary Turing machine TM with a polynomial bound $P(n)$ accepting the same language as TM_1. (In general, $P(n)$ assumes greater values than $P_1(n)$ but is still a polynomial.) Similarly, every language that is polynomially bounded with respect to any standard Turing machine model or with respect to any reasonable model of computation belongs to the family $\mathcal{P}$, as defined above with respect to the ordinary Turing machine.

The family $\mathcal{P}$ is also of crucial importance because languages outside $\mathcal{P}$ can be visualized as impossible to compute. In fact, we say that a recursive language is **intractable** if it does not belong to $\mathcal{P}$.

Clearly, languages outside $\mathcal{P}$ are intractable from the practical point of view. (See also Exercise 6.8.) The same can be said about such languages in $\mathcal{P}$, where the polynomial bound is a huge one. However, it would not be natural to draw the borderline between tractability and intractability somewhere inside $\mathcal{P}$. Such a definition would also be time-varying: Drastic developments in computers could change it. On the other hand, the family $\mathcal{P}$ provides a very natural characterization of tractability.

We now consider **nondeterministic** Turing machines: When scanning a specific symbol in a specific state, the machine may have several possibilities for its behavior. Otherwise, a nondeterministic machine is defined as a deterministic one. A word w is accepted iff it gives rise to an accepting computation, independently of the fact that it might also give rise to computations leading to failure. Thus, as in connection with nondeterministic machines in general, all roads to failure are disregarded if there is one possible road to success.

The time required by a nondeterministic Turing machine TM to accept a word $w \in L(TM)$ is defined to be the number of steps in the shortest computation of TM accepting w. The **time-complexity function** of TM is now defined by

$$t_{TM}(n) = \max\{1, m \mid TM \text{ accepts in time } m \text{ some } w \text{ where } |w| = n\}.$$

Thus, only accepting computations enter the definition of $t_{TM}(n)$. If no words of length n are accepted, we obtain $t_{TM}(n) = 1$.

We omit further formal details of the definition of a nondeterministic Turing machine. The reader might want to study how the formal definition presented in Chapter 4 should be changed in case of nondeterministic machines.

Having defined the time-complexity function, we now define the notion of a **polynomially bounded nondeterministic** Turing machine exactly as before. We denote by $\mathfrak{N}\mathcal{P}$ the family of languages acceptable by nondeterministic polynomially bounded Turing machines.

Problems in $\mathcal{P}$ are tractable, whereas problems in $\mathfrak{N}\mathcal{P}$ have the property that it is tractable to check whether or not a good guess for the solution of the problem is correct. A nondeterministic Turing machine may be visualized as a device that checks whether or not a guess is correct: It makes a guess (or several guesses) at some stage during its computation, and the final outcome is acceptance only in case the guess was (or the guesses were) correct. Thus, a time bound for a nondeterministic Turing machine is, in fact, a time bound for checking whether or not a guess for the solution is correct.

The reader should have no difficulties in verifying that the languages SAT (as well as its modifications) and KNAPSACK considered in Examples 6.1 and 6.2 are in $\mathfrak{N}\mathcal{P}$. Consider, for instance, the language SAT. A nondeterministic Turing machine *TM* guesses the value 1 or 0 for each variable and remembers (this is done easiest on a separate tape) the variables for which the value has already been assigned. The evaluation of the entire wffpc follows the customary procedure. It is obvious that polynomial time will suffice.

Indeed, the following stronger result can be obtained.

Theorem 6.15. *The language 2-CONSAT is in $\mathcal{P}$.*

Proof. Consider the algorithm presented in Example 6.1. Clearly, each of Steps 1–3 is polynomially bounded with regard to time. The wffpc α''' resulting from Step 3 may be longer than α. However, a polynomial upper bound for its length in terms of the length of α is easy to obtain. Since the number of variables is properly reduced before every return to Step 1, a polynomial time bound for processing α results. The construction of explicit bounds is left to the reader in Exercise 6.9. It should be observed that the format of the input is quite irrelevant: A polynomial time bound is obtained independently of the fact of whether the input is an arbitrary word or a wffpc in 2-conjunctive normal form. This follows because the checking of a proper format (such as 2-conjunctive normal form) can be done quickly by a Turing machine. □

Clearly, $\mathcal{P}$ is contained in $\mathfrak{N}\mathcal{P}$. However, it is not known whether or not the containment is proper. The problem of whether or not $\mathcal{P}$ equals $\mathfrak{N}\mathcal{P}$ can justly be called the most celebrated open problem in the theory of computation. The significance of this question is due to the fact that many practically important problems are known to be in $\mathfrak{N}\mathcal{P}$, whereas it is not known whether or not they are in $\mathcal{P}$. In fact, all known deterministic algorithms for these problems are exponential as far as time is concerned. Thus, a proof of $\mathcal{P} = \mathfrak{N}\mathcal{P}$ would make all of these problems tractable.

Nondeterministic Turing machines and guessing are not, as such, intended to be modeling computation. Nondeterminism is merely an auxiliary notion and, as we shall see, a very useful one. Indeed, if the reader wants to settle the question of whether or not $\mathcal{P} = \mathfrak{N}\mathcal{P}$, the following definitions and results show that it suffices to consider one particular language (might be a favorite one!) and determine whether or not it is in $\mathcal{P}$. There is a great variety of such $\mathfrak{N}\mathcal{P}$-complete languages, resulting from practically all areas of mathematics. We now define $\mathfrak{N}\mathcal{P}$-completeness.

Definition 6.3. *A language $L_1 \subseteq \Sigma_1{}^*$ is **polynomially reducible** to a language $L_2 \subseteq \Sigma_2{}^*$—in symbols, $L_1 \leq_P L_2$—iff there is a deterministic polynomially bounded Turing machine TM translating words w_1 over Σ_1 into words w_2 over Σ_2 in such a way that w_1 being in L_1 is equivalent to its image w_2 being in L_2.*   □

Observe that the Turing machine *TM* introduced in the previous definition must halt with all inputs—this is a consequence of *TM* being deterministic and polynomially bounded. The relation $\leq_P$ resembles the reducibilities considered in Chapter 4. In particular, it is easy to see that $\leq_P$ is transitive. (One considers the "composition" of the two polynomially bounded Turing machines involved.) Hence, if we define

$$L_1 \equiv_P L_2 \quad \text{iff} \quad L_1 \leq_P L_2 \quad \text{and} \quad L_2 \leq_P L_1,$$

the relation $\equiv_P$ thus obtained is an equivalence relation. Two languages L_1 and L_2 are called **polynomially equivalent** iff $L_1 \equiv_P L_2$ holds. Similar matters are also discussed in Exercise 6.10.

The next result is an immediate consequence of the definition.

Lemma 6.16. *Whenever $L_1 \leq_P L_2$ and L_2 is in $\mathcal{P}$, then L_1 is also in $\mathcal{P}$.*

Definition 6.4. *A language L is $\mathcal{NP}$-hard iff $L' \leq_P L$ holds for every language L' in $\mathcal{NP}$. L is termed $\mathcal{NP}$-complete iff it is $\mathcal{NP}$-hard and, in addition, belongs to $\mathcal{NP}$.*   □

$\mathcal{NP}$-complete languages can be visualized to represent the hardest problems in $\mathcal{NP}$. Moreover, to settle the question of whether or not $\mathcal{P} = \mathcal{NP}$, it suffices to decide whether or not an arbitrary $\mathcal{NP}$-complete language L is in $\mathcal{P}$. Indeed, consider such a language L. If L is not in $\mathcal{P}$, then clearly $\mathcal{P} \neq \mathcal{NP}$. If L is in $\mathcal{P}$, then the definition of $\mathcal{NP}$-completeness and Lemma 6.16 show that every language in $\mathcal{NP}$ is also in $\mathcal{P}$. But this means that $\mathcal{P} = \mathcal{NP}$.

As customary in similar contexts, the definition of polynomial reducibility means that the Turing machine, *TM*, can be effectively constructed from L_1 and L_2. Hence, an algorithm operating in polynomial time can be constructed for every problem in $\mathcal{NP}$, provided a (deterministic) polynomial-time-bounded algorithm is known for some $\mathcal{NP}$-complete problem. (We speak occasionally of problems instead of languages to remind the reader of the interchangeability of the notions.) Thus, once you get a polynomial-time-bounded algorithm for one of the very many $\mathcal{NP}$-complete problems, you get a polynomial-time-bounded algorithm for every problem in $\mathcal{NP}$! In view of the vast effort spent in trying to improve the known algorithms for some of such problems (because of their great practical significance) and in view of the fact that such an effort has not led to success, it is now generally believed that $\mathcal{P} \neq \mathfrak{N}\mathcal{P}$.

The following lemma is an immediate consequence of the transitivity of the relation $\leq_P$.

Lemma 6.17. *If L_1 is $\mathfrak{N}\mathcal{P}$-complete and L_2 is a language in $\mathfrak{N}\mathcal{P}$ satisfying $L_1 \leq_P L_2$, then the language L_2 is also $\mathfrak{N}\mathcal{P}$-complete.*

Lemma 6.17 is a basic tool in showing that a specific language is $\mathfrak{N}\mathcal{P}$-complete. However, in order to apply Lemma 6.17, we need something with which to start—that is, we need one $\mathfrak{N}\mathcal{P}$-complete language. Such a language will now be exhibited.

Theorem 6.18. *The language SAT is $\mathfrak{N}\mathcal{P}$-complete.*

Proof. We have already observed that SAT is in $\mathfrak{N}\mathcal{P}$. Hence, it suffices to prove that every language L in $\mathfrak{N}\mathcal{P}$ satisfies $L \leq_P \text{SAT}$.

Consider an arbitrary language $L \subseteq \Sigma^*$ accepted by a nondeterministic polynomially bounded Turing machine, *TM*. Let $P(n)$ be the polynomial in question. Without loss of generality, we assume that $P(n) \geq n$ holds for all n. To prove that $L \leq_P \text{SAT}$, we construct an algorithm transforming an arbitrary word w over Σ into a wffpc α such that α is satisfiable iff w is in L. Moreover, the algorithm will be time-bounded by a polynomial $Q(n)$ (that depends on *TM* and $P(n)$). Since the informal algorithm can be replaced by a Turing machine with a polynomial time bound $Q'(n)$, the relation $L \leq_P \text{SAT}$ follows. (The bound $Q'(n)$ might assume greater values than $Q(n)$ but is still a polynomial in n.)

Thus, consider an arbitrary word $w \in \Sigma^*$. Let $|w| = n$. We begin now the construction of a wffpc α such that α is satisfiable iff w is in L. The wffpc α will not be strictly in conjunctive normal form. To improve readability, we make use of the implication sign, $\rightarrow$. Thus, $\beta_1 \rightarrow \beta_2$ stands for $\sim\beta_1 \vee \beta_2$. The wffpc α will be a conjunction of certain wffpc's α_i, but the wffpc's α_i are not necessarily disjunctions of literals. However, we still refer to the wffpc's α_i as **clauses** and to a conjunction of some of the α_i's as a **group of clauses**.

The word w is in L iff there is a computation C of *TM*, starting from the initial configuration with w as the input and leading to acceptance. Moreover, the length of C is at most $P(n)$: We have to consider what happens only at the time instants $0, 1, \ldots, P(n)$. It is no loss of generality to assume that *TM* continues operating (for example, without doing anything) after eventually reaching the accepting state. In this fashion we may assume that the "life" of *TM* always continues until the time instant $P(n)$—a very helpful assumption from the point of view of certain technicalities in the sequel.

The time bound $P(n)$ also shows that at most $P(n) + 1$ squares of the tape are visited in the course of the computation C. In what follows, we denote by (i) the initial condition:

(i) At time instant 0, the word w is written in the squares numbered $0, 1, \ldots, n-1$, whereas the squares numbered $n, \ldots, P(n)$ and $-1, -2, \ldots, -P(n)$ are provided with the blank symbol. Moreover, TM is scanning the square 0 in the initial state.

The squares numbered $-1, -2, \ldots$ are, of course, to the left from the initial square. We are thus using $2P(n) + 1$ squares in the sequel because we do not want to exclude the possibility of TM moving to the left from the initial square. (In fact, we also could exclude this possibility, since everything discussed here remains unchanged even if TM can extend its tape to the right only.)

We also consider the following final condition:

(ii) At time instant $P(n)$, TM is in the final state.

For every s and t with

$$-P(n) \leq s \leq P(n) \quad \text{and} \quad 0 \leq t \leq P(n),$$

we say that the square s is in the **state** $(a, 0)$ (resp. (a, q)) at time instant t iff the tape symbol a appears in s at t and, moreover, s is unscanned at t (resp. s is scanned at t by TM in the state q).

Conditions (iii) and (iv) express the fact that the states of the squares can be changed only according to the rules of TM.

(iii) Assume that $-P(n) \leq s \leq P(n)$ and $0 \leq t \leq P(n) - 1$ and that the square s is at time instant t in the state $(a, 0)$. Then s is at time instant $t + 1$ either in the state $(a, 0)$ or in one of the states described by (*) and (**): (*) In the state (a, q'), provided $s \leq P(n) - 1$, the square $s + 1$ is at t in the state (b, q), and TM has an instruction to go left and change to state q' when scanning b in the state q. (**) In the state (a, q''), provided $s \geq -P(n) + 1$, the square $s - 1$ is at t in the state (b, q), and TM has an instruction to go right and change to q'' when scanning b in the state q.

(iv) Assume that $-P(n) \leq s \leq P(n)$ and $0 \leq t \leq P(n) - 1$ and that the square s is at time instant t in the state (a, q). Then s is in one of the states described by (*) and (**) at time instant $t + 1$: * In the state (b, q'), provided TM has an instruction to write b and change to state q' when scanning a in the state q. (**) In the state $(a, 0)$, provided it is possible for TM to go either right or left when scanning a in the state q.

If our Turing machine, TM, were deterministic, we could say that w is in L iff each of the conditions (i)–(iv) is satisfied. Since TM is nondeterministic, we have to introduce the following additional condition in order to avoid situations where, for instance, condition (iii) is used to get two squares entering some q-state at time instant $t + 1$:

(v) For every s and t in their proper ranges, there is exactly one pair (a, x) such that the square s is at t in the state (a, x). Moreover, for every t, there is exactly one s such that in the pair (a, x) we have $x \neq 0$ (that is, x is one of the state symbols q of TM).

It is now clear that w is in L iff each of conditions (i)–(v) is satisfied. We still have to express the conjunction of conditions (i)–(v) in terms of a wffpc α. But this is going to be fairly easy. The variables of α will be of the form

$$[s, t, a, x], \tag{36}$$

where $-P(n) \leq s \leq P(n)$, $0 \leq t \leq P(n)$, a ranges over the tape symbols of TM, and x ranges over the states of TM and over the special symbol 0. The intuitive meaning of a variable form (36) is: The square s is at time instant t in the state (a, x). The number of the variables in (36) is of the order $(P(n))^2$; that is, the number is less than or equal to

$$c(P(n))^2,$$

where c is a constant independent of n.

If B is the blank symbol, q_0 is the initial state of TM, and $w = a_1 \cdots a_n$, where the a's are letters, then condition (i) can be expressed as the conjunction of the literals

$$[-P(n), 0, B, 0], \ldots, [-1, 0, B, 0], [0, 0, a_1, q_0],$$
$$[1, 0, a_2, 0], \ldots, [n - 1, 0, a_n, 0], [n, 0, B, 0], \ldots,$$
$$[P(n), 0, B, 0].$$

The length of the conjunction is of the order $P(n)$, and the conjunction can be written down in approximately $P(n)$ steps. These estimates should be multipled by $\log n$ if we take into account the fact that the notation of the variables is dyadic. In any case, the group of clauses corresponding to condition (i) can be written down in the order of $nP(n)$ steps.

Condition (ii) is expressed as the disjunction of the literals

$$[s, P(n), a, q_F],$$

where $-P(n) \leq s \leq P(n)$, a ranges over the tape symbols of TM, and q_F is the final state of TM. Thus, condition (ii) is represented by only one clause in α. The length of the clause is of the order $P(n)$, and the clause can be written down in the order of $nP(n)$ steps.

Condition (iii) can be expressed as a group of clauses, each clause being of the form

$$[s, t, a, 0] \to ([s, t + 1, a, 0] \lor D), \tag{37}$$

where D is a disjunction of terms, each of which is a conjunction of two variables. The length of the clauses is bounded by a constant independent of n. The number of the clauses is of the order $(P(n))^2$, and so the whole group of clauses can be written down, by directly translating (iii) into conditions of the form (37), in the order of $n(P(n))^2$ steps.

The group of clauses corresponding to condition (iv) consists of clauses of the form

$$[s, t, a, q] \rightarrow D,$$

where D is a disjunction of variables. Again the length of the individual clauses is bounded by a constant independent of n and the whole group of clauses can be written down in the order of $n(P(n))^2$ steps.

Consider, finally, condition (v). For fixed s and t, the first sentence of (v) can be expressed in the form

$$D \wedge C, \tag{38}$$

where D is the disjunction of all variables $[s, t, a, x]$ such that a and x assume all their appropriate values, and C is the conjunction of terms of the form $\sim y_i \vee \sim y_j$, where the y's range over the letters appearing in D. Thus, the length of (38) is bounded by a constant independent of n and, consequently, we obtain the upper bound $n(P(n))^2$ for the number of steps needed to write down the group of clauses expressing the first sentence of (v), exactly as before.

Consider the second sentence of (v). For a fixed t, we first express the requirement that at least one square is in some q-state at t. This will be a disjunction of variables, the number of variables in the disjunction being of the order $P(n)$. When this is repeated for all possible values of t, we obtain a group of clauses with (the order of) $P(n)$ clauses. We still have to express the requirement that there is no t with the property that two squares are in a q-state at t. Considering how a square can possibly be activated (that is, changed from 0-state to a q-state), we observe that it suffices to express the following requirement: Whenever a square s is at t in some q-state, then both of the squares $s + 1$ and $s + 2$ are at t in 0-state. Thus, we obtain a conjunction of implications of the form

$$[s, t, a, q] \rightarrow D_1 \wedge D_2, \tag{39}$$

where D_1 (resp. D_2) is the disjunction of letters $[s + 1, t, b, 0]$ (resp. $[s + 2, t, b, 0]$) with b ranging over the tape symbols of TM. The number of implications in (39) is of the order of $(P(n))^2$, whereas the length of each implication is bounded by a constant independent of n. But this means that the whole group of clauses corresponding to condition (v) can be written down in the order of $n(P(n))^2$ steps.

Therefore, in the order of $n(P(n))^2$ steps (and, hence, in polynomial time), one can construct a wffpc α such that w is in L iff α is satisfiable. $\square$

Slightly stronger versions of Theorem 6.18 will be presented in our next two theorems. In many reduction arguments, Theorems 6.19 and 6.20 are easier to apply than the original Theorem 6.18.

Theorem 6.19. *The language CONSAT is $\mathfrak{N}\mathcal{P}$-complete.*

Proof.   Our proof is a direct continuation of the proof of Theorem 6.18: We transform the wffpc α into a wffpc α' in conjunctive normal form.

We assume that the implications $\beta_1 \rightarrow \beta_2$ appearing in α have been written in the form $\sim\beta_1 \vee \beta_2$. This can be done when writing down α, and it does not affect the time complexity estimate obtained for α in the proof of Theorem 6.18. Furthermore, the distributive law is applied for $D_1 \wedge D_2$ appearing in implications (39). After these modifications, the wffpc α is "almost" in conjunctive normal form, the only possible exception being clauses

$$(x_1 \vee \cdots \vee x_t \vee y_1z_1 \vee \cdots \vee y_uz_u), \tag{40}$$

where x's, y's, and z's are literals. (Conjunction is denoted here simply by juxtaposition.) The exceptional clauses may result from condition (iii) or the second sentence of condition (v). (In fact, a more-detailed analysis could reveal further facts, although this is not very essential at this point. For instance, in the exceptional clauses, we have always $t \leq 2$ and we can also say that all variables, with the possible exception of x_1, are unnegated.) Moreover, u is a constant independent of n.

But now each clause in (40) can be replaced by the conjunction of all terms of the form

$$(x_1 \vee \cdots \vee x_t \vee v_1 \vee \cdots \vee v_u), \tag{41}$$

where $v_i = y_i$ or $v_i = z_i$ for $i = 1, \ldots, u$. The satisfiability is not affected by this replacement: In fact, the new wffpc will be equivalent to the old one with respect to any truth-value assignment. Of course, the number of terms (41) will be 2^u, but because u is independent of n, we are able to write down the new wffpc α' (that is in conjunctive normal form) in the order of $n(P(n))^2$ steps, exactly as before. $\square$

Theorem 6.20. *The language 3-CONSAT is $\mathfrak{N}\mathcal{P}$-complete.*

Proof.   By the previous theorem, it suffices to observe that a clause

$$x_1 \vee x_2 \vee \cdots \vee x_t, \qquad t \geq 4,$$

in conjunctive normal form can be replaced by the wffpc

$$(x_1 \vee x_2 \vee y_1) \wedge (x_3 \vee \sim y_1 \vee y_2) \wedge (x_4 \vee \sim y_2 \vee y_3) \wedge \cdots$$
$$\wedge (x_{t-2} \vee \sim y_{t-4} \vee y_{t-3}) \wedge (x_{t-1} \vee x_t \vee \sim y_{t-3}),$$

where $y_1, \ldots, y_{t-3}$ are new variables. $\square$

The reader is at this point reminded of Theorem 6.15. We see that there is an essential difference between 2-conjunctive and 3-conjunctive normal forms, although the straightforward truth-table technique yields an exponential algorithm in both cases.

The next theorem gives an example about establishing $\mathfrak{N}\mathcal{P}$-completeness by the reduction technique. Some further examples are contained in Exercises 6.11–6.13. The reader is referred to [GaJ] for a comprehensive exposition on $\mathfrak{N}\mathcal{P}$-complete problems of various types.

Theorem 6.21. *KNAPSACK is $\mathfrak{N}\mathcal{P}$-complete.*

Proof. We shall show that

$$3\text{-CONSAT} \leq_P \text{KNAPSACK}. \tag{42}$$

This implies, by Lemma 6.17 and Theorem 6.20, that KNAPSACK is $\mathfrak{N}\mathcal{P}$-complete.

Relation (42) is established by giving an algorithm that will transform a given wffpc α into a knapsack problem β such that β has a solution iff α is satisfiable. It will be obvious that the algorithm operates in polynomial time.

Without loss of generality, we assume that α is in 3-conjunctive normal form:

$$\alpha = \alpha_1 \wedge \cdots \wedge \alpha_n,$$

where each α_i is a disjunction of at most three of the literals $x_1, \ldots, x_m, \sim x_1, \ldots, \sim x_m$. This is no loss of generality because we can check (very quickly!) whether or not the original α is in 3-conjunctive normal form and, if it is not, transform α into a knapsack problem having (trivially) no solutions. The number n can be viewed as the size of α: The number of all symbols in α is less than or equal to $10n$. (Here we count variables as single symbols.) We also observe at this point that it makes no difference if the numbers are in decimal notation in the knapsack problem β we are constructing. This follows because we can certainly change the decimal notation into the dyadic one in polynomial time.

Thus, we have to define the integers w_0 (size of the knapsack) and $w_1, \ldots w_k$ (sizes of the individual items). The integers will be given in decimal notation. By definition,

$$w_0 = \underbrace{3 \cdots 3}_{m+n}.$$

By definition, $k = 2m + 2n$. We denote the numbers $w_1, \ldots, w_k$ by

$$x_1, \ldots, x_m, \qquad \sim x_1, \ldots, \sim x_m, \qquad \alpha_1, \ldots, \alpha_n, \qquad \alpha_1', \ldots, \alpha_n',$$

indicating how the numbers are associated to the items introduced above. All these numbers will have $m + n$ digits, just as w_0.

For $i = 1, \ldots, m$, the ith digit from the right in x_i (resp. in $\sim x_i$) equals 3 and, for $j = 1, \ldots, n$, the $(m + j)$th digit from the right in x_i (resp. $\sim x_i$) equals 1 if x_i (resp. $\sim x_i$) occurs in α_j. All other digits in x_i and $\sim x_i$ equal 0.

For $j = 1, \ldots, n$, the $(m + j)$th digit from the right in α_j equals 1, all other digits in α_j being equal to 0. By definition, $\alpha_j' = \alpha_j$ for $j = 1, \ldots, n$. This completes the definition of the knapsack problem β.

As an illustration, consider the wffpc of (34) discussed at the beginning of this section. In this case, we have $n = 4$ and $m = 3$. The knapsack problem β is in this case defined by the following integers (in decimal notation):

$$
\begin{aligned}
w_0 &= 3333333, & x_1 &= 0001003, \\
\sim x_1 &= 0110003, & x_2 &= 0010030, \\
\sim x_2 &= 0101030, & x_3 &= 1000300, \\
\sim x_3 &= 0000300, & \alpha_1 &= \alpha_1' = 0001000, \\
\alpha_2 &= \alpha_2' = 0010000, & \alpha_3 &= \alpha_3' = 0100000, \\
\alpha_4 &= \alpha_4' = 1000000.
\end{aligned}
$$

It is immediately verified that

$$
w_0 = \sim x_1 + \sim x_2 + x_3 + \alpha_1 + \alpha_1' + \alpha_2 + \alpha_2' + \alpha_3 + \alpha_4 + \alpha_4'
$$

and, consequently, the knapsack problem β has a solution. The key part of the solution is $\sim x_1 + \sim x_2 + x_3$. It indicates the truth-value assignment showing that (34) is satisfiable: x_1 and x_2 are assigned the value 0, whereas x_3 is assigned the value 1. The α's are needed in the solution to "fill up" eventual gaps in the first n digits from the left. In general, each of the first n digits equals 1, 2, or 3 in the key part, depending on whether 1, 2, or 3 chosen literals occur in the disjunction in question. (Since we are dealing with a 3-conjunctive normal form, no more than three literals may occur.) Thus, either α_j or both α_j and α_j' might be needed.

The reader might want to construct the knapsack problem resulting from the wffpc in (35) and verify that it has no solution.

At this stage it should also be fairly clear that the construction presented above works: the knapsack problem β resulting from the given wffpc α possesses a solution iff α is satisfiable.

Indeed, assume that α is satisfiable. For each $i = 1, \ldots, m$, we choose either x_i or $\sim x_i$ to the solution of β, depending on whether the variable x_i is assigned the value 1 or 0 in the assignment giving α the value 1. In the sum thus obtained, the last m digits equal 3, whereas each of the first n digits equals 1, 2, or 3. The last assertion follows because at least one of the chosen literals occurs in each of the disjunctions α_j. Thus, by a suitable set of the α-numbers, it is possible to increase the entire sum in such a way that it equals w_0.

Conversely, assume that β possesses a solution. This means that there is a subset S of the set of numbers

$$
x_1, \ldots, x_m, \qquad \sim x_1, \ldots, \sim x_m, \qquad \alpha_1, \ldots, \alpha_n, \qquad \alpha_1', \ldots, \alpha_n'
$$

such that the sum of the numbers in S equals $w_0 = 3\ldots3$. (Every number may occur in the solution at most once!) Hence, for $i = 1,\ldots, m$, exactly one of x_i and $\sim x_i$ occurs in S. Otherwise, one of the last m digits in the sum differs from 3. Moreover, the numbers occurring in S indicate the truth-value assignment giving α the value 1. This follows because every disjunction of α must contain at least one of the corresponding literals. Otherwise, one of the first n digits in the sum is less than or equal to 2. Therefore, α is satisfiable. $\square$

We have already pointed out that many problems resulting from diverse areas of mathematics are known to be $\mathfrak{N}\mathcal{P}$-complete. However, all known $\mathfrak{N}\mathcal{P}$-complete problems have certain far-reaching similarities. They are all, when viewed as languages, polynomially equivalent in the sense defined above. Even more can be said: All known $\mathfrak{N}\mathcal{P}$-complete languages are **polynomially isomorphic**; that is, between any two known $\mathfrak{N}\mathcal{P}$-complete languages, there are polynomial-time one-to-one and onto mappings with polynomial-time inverses. It has also been shown that if all $\mathfrak{N}\mathcal{P}$-complete languages are actually polynomially isomorphic, then $\mathcal{P} \neq \mathfrak{N}\mathcal{P}$.

Let us call a language L **sparse** iff there is a polynomial $P(n)$ such that the number of words in L of length at most n is less than or equal to $P(n)$. (The term *sparse* reflects the fact that, in general, the number of all words of length n grows exponentially with n.) For instance, the language SAT is not sparse. No sparse $\mathfrak{N}\mathcal{P}$-complete language is known. In fact, if one of two polynomially isomorphic languages is sparse, then so is the other.

It has been shown that if there exists a sparse $\mathfrak{N}\mathcal{P}$-hard language, then $\mathcal{P} = \mathfrak{N}\mathcal{P}$. From this it easily follows that determining the existence of a sparse $\mathfrak{N}\mathcal{P}$-complete language is equivalent to solving the $\mathcal{P} = \mathfrak{N}\mathcal{P}$ problem.

We mention, finally, some results dealing with Diophantine sets. By Theorem 5.9, for every recursively enumerable set S, there is a polynomial P such that, for all x,

$$x \in S \quad \text{iff} \quad (\exists y_1 \geq 0)\ldots(\exists y_m \geq 0)[P(x, y_1, \ldots, y_m) = 0].$$

A natural problem is to try to characterize $\mathfrak{N}\mathcal{P}$ in terms of similar representations—for instance, by imposing restrictions on the quantifiers. (As natural in connection with Diophantine representations, we view here elements of $\mathfrak{N}\mathcal{P}$ as sets of numbers, obtained from languages by some polynomial-time encoding.) In this fashion, interconnections between Turing machine complexity and the complexity of the Diophantine representations can be obtained. For instance, if the range of the existentially quantified variables y_i is bounded from above by a polynomial in x, then the resulting set is in $\mathfrak{N}\mathcal{P}$. Every set in $\mathfrak{N}\mathcal{P}$ has a representation where the range mentioned is bounded from above by a function $2^{Q(x)}$, where Q is a polynomial.

These two results do not give an exhaustive characterization of $\mathfrak{N}\mathcal{P}$ in terms of a necessary and sufficient condition. Such a characterization has been recently given in [KeH], using both bounded existential and bounded universal quantifications.

6.5. PROVABLY INTRACTABLE PROBLEMS

Although a problem is decidable, it might still be intractable in the sense that any algorithm for solving it must use unmanageable amounts of computational resources. In the previous section, intractability was defined formally by the condition of being outside $\mathcal{P}$.

We have also observed that, for very many important problems, all algorithms presently known operate in exponential time, whereas it has not been shown that these problems actually are intractable. All $\mathcal{NP}$-complete problems belong to this class. Indeed, establishing *lower bounds* is generally a difficult task in complexity theory. The complexity of a particular algorithm can usually be estimated, but it is harder to say something general about all algorithms for a particular problem—for instance, to give a lower bound for their complexity. Results such as the speedup theorem show that this might even be impossible in some cases. The reader is also reminded of the fact that according to the theory of abstract complexity measures developed earlier, we cannot expect to prove that any specific problem is intrinsically difficult in all measures. However, there are also a number of problems known to be intractable according to the definition of intractability given in the previous section. While some of these problems might seem artificial because they are constructed in such a way that they require an abundance of some specific resource such as time or space, the following one is very natural.

Regular expressions were defined in Section 3.2. We now consider **regular expressions with squaring.** This means that the following clause is added to Definition 3.3 given in Section 3.2: If w is a regular expression over Σ, then so is w^2. The regular expression w^2 denotes the square L^2 of the language L denoted by w.

It is clear that this addition to the definition of a regular expression does not change the family DRE of languages denoted by regular expressions. However, the addition provides more-succinct representations for some languages, which is an important issue in the sequel.

It is obviously decidable whether or not the complement of a language denoted by a regular expression with squaring is empty. However, we shall see that this problem is not in $\mathcal{P}$. In fact, it can be shown that exponential space is needed for the solution. Hence, exponential time is also needed.

The idea in the proof that exponential space is needed is the following. Consider a Turing machine TM using at most exponential space. (For instance, TM uses no more than 2^n squares when processing an input of length n.) For a word w with length n, we write a regular expression with squaring denoting the language consisting of all words that are *not* computations of TM for the input w. Thus, the complement of this language is empty iff w is not in $L(TM)$. The regular expression denoting the language of "noncomputations" is of length proportional to n. We then choose TM in such a way that any Turing machine accepting $L(TM)$ uses essentially exponential space. Since a word is in $L(TM)$ iff the complement of the language denoted by the constructed regular expression with squaring is nonempty, the decision of the emptiness of the complement of such a language cannot be accomplished in much less space.

We now present more details of the idea described above. The whole argument given is still only an outline. However, all missing details are similar to matters already discussed—some of them a number of times!—and so the reader should not have any difficulties with them.

We omit the formal details of the definition of the space-complexity function of a Turing machine, as well as related notions, such as a Turing machine being *polynomially space-bounded*. All definitions are analogous to the ones dealing with time complexity. It is obvious that if a language is not acceptable by any polynomially space-bounded Turing machine, then the language is not in $\mathcal{P}$.

The Turing machines considered in this section will be deterministic. We shall assume, however, that the model is modified in such a way that the tape is never extended to the left. This is not essential but will be convenient in the sequel and does not affect any of the issues discussed.

Let TM be a Turing machine with the space complexity function 2^n. As in our earlier proofs, we consider letters of the form $[a, x]$, where a is a tape symbol of TM and x is either a state of TM or the symbol 0. Again the intuitive meaning is: The symbol a is either scanned by TM in the state q or else unscanned. Let Σ be the alphabet consisting of all letters $[a, x]$ and, in addition, of the special letter $\Rightarrow$.

Then an input word $w = w_1 \cdots w_n$, where each w_i is a letter, is accepted by TM iff there is a computation of TM of the form

$$[w_1, q_0][w_2, 0] \cdots [w_n, 0][B, 0]^{2^n} \Rightarrow \beta_1 \Rightarrow \cdots \Rightarrow \beta_k, \tag{43}$$

where each β_i results from β_{i-1} (and β_1 results from the initial part) by the instructions of TM, and β_k contains the final state—that is, a letter of the form $[a, q_F]$.

Thus, in (43) we already initially provide all the space TM can possibly need when processing w. Actually $2^n - n$ blanks would suffice, but the number 2^n is more convenient to handle in connection with the squaring operation. Of course, k may be very large in comparison with n.

We now write down a regular expression with squaring, say α, over the alphabet Σ such that the language L_α denoted by α consists of all words over Σ that are *not* of the form of (43). Thus, w is in $L(TM)$ iff

$$L_\alpha \neq \Sigma^*.$$

Consequently, deciding the emptiness of the complement of L_α amounts to deciding whether or not w is in $L(TM)$.

The length of α will be proportional to n—that is, $|\alpha| \le cn$, where c is a constant (depending on TM). To achieve this, the squaring operation is quite essential. For instance, $[B, 0]^{2^n}$ is obtained by applying squaring n times to $[B, 0]$; that is, denoted by a regular expression of length $n + 1$.

Clearly, a word x over the alphabet Σ is not of the form (43) iff x satisfies at least one of conditions (i)–(iii):

(i) x begins wrong—that is, x does not have the word

$$[w_1, q_0][w_2, 0] \cdots [w_n, 0][B, 0]^{2^n} \Rightarrow$$

as its prefix.

(ii) x ends wrong—that is, β_k contains no letter $[a, q_F]$.

(iii) x does not depict a computation of TM—that is, there are two consecutive parts of x between symbols $\Rightarrow$ such that the latter does not result from the former by the instructions of TM.

The regular expression α is the union of three regular expressions (with squaring) α_a, α_b, and α_c describing conditions (i)–(iii), respectively. The length of each of the regular expressions α_a, α_b, and α_c must be proportional to n.

The regular expression α_a is obtained by a direct translation of the following condition: Either the first symbol of the word x is wrong (translation: $(\Sigma - [w_1, q_0])\Sigma^*$), or else the first symbol is correct but the second is wrong, or else the two first symbols are correct but the third is wrong, and so on, or else the n first symbols are correct but there is some symbol other than $[B, 0]$ before the first appearance of the symbol $\Rightarrow$, or else the first n symbols are correct and only symbols $[B, 0]$ occur before the first appearance of $\Rightarrow$, but there are too many or too few of them.

The translation of the part of the condition dealing with symbols $[B, 0]$ requires the squaring operation in order to keep the length of α_a proportional to n.

Writing down α_b is very easy: We just say that the word x does not contain any letter $[a, q_F]$. Squaring is not needed in writing down this regular expression, and the length of α_b will be independent of n.

The construction of the regular expression α_c is based on the following intuitive idea. A ruler with two holes, each of length three and each at a proper distance from the other, is at our disposal. We "read" words using this ruler, looking for "improper" pairs of words of length three.

The whole regular expression α_c is a union consisting of the following terms. For a specific letter $[a, q]$, assume that TM has the instruction to go to q' and write b when scanning a in the state q. Then the corresponding term in the union defining α_c is

$$\Sigma^*[a, q]\Sigma^{2^n}\Sigma^n(\Sigma - [b, q'])\Sigma^*.$$

Similarly, if *TM* has the instruction of going to the left and entering the state q' when scanning a in the state q, then the corresponding term in the union defining α_c is

$$\cup \Sigma^*[b, 0][a, q]\Sigma^{2^n}\Sigma^{n-1}((\Sigma - [b, q'])\Sigma^* \cup \Sigma(\Sigma - [a, 0])\Sigma^*)$$

where this particular union ranges over all symbols $[b, 0]$ such that b belongs to the tape alphabet of *TM*. The right-instructions of *TM* are handled similarly.

We still express the fact that a symbol can be changed only by *TM*. This is basically a union

$$\cup \Sigma^* c_1 c_2 c_3 \Sigma^{2^n}\Sigma^{n-1}(\Sigma - c_2)\Sigma^*$$

over triples (c_1, c_2, c_3), where c_2 ranges over all symbols not of form $[a, q]$, and the ranges of c_1 and c_3 include the same symbols and, in addition, properly chosen symbols $[a, q]$. For instance, $[a, q]$ belongs to the range of c_1 if the instruction for *TM* for the pair (a, q) is not "move right." Moreover, the symbol $\Rightarrow$ is never introduced or eliminated.

We have now completed the construction of the regular expression α in such a way that $|\alpha|$ will be proportional to n.

Finally, we assume that the Turing machine, *TM*, discussed above actually satisfies also the condition of Exercise 6.16: The language $L(TM)$ requires essentially $2^{n/2}$ space. If there were a Turing machine deciding, in polynomial space, the emptiness of the complement of the language denoted by a regular expression with squaring, we would obtain a Turing machine deciding membership in $L(TM)$ in polynomial space. This is a contradiction for large enough values of n. Hence, we have established the following result.

Theorem 6.22. *The problem of deciding whether or not the complement of the language denoted by a regular expression with squaring is empty is not polynomially space-bounded. Hence, this problem is not in* $\mathcal{P}$.

Theorem 6.22 remains valid if, instead of squaring, we allow intersection as an additional operation when defining regular expressions. If both intersection and complementation are allowed as additional operations, then the same problem (deciding the emptiness of the complement) becomes even more intractable: It is not space-bounded by any function of the form

$$f(n) = 2^{2^{2^{\cdot^{\cdot^{2^n}}}}},$$

where the stack of exponents is of length m, for some m.

It has been known since the famous work of Gödel in 1931 that formalized arithmetic is undecidable. However, it can be shown (although the proof is not easy) that the theory of addition is actually decidable. The theory of addition, often referred to as *Presburger arithmetic,* consists of first-order logic and the standard axioms for addition. It can also be shown that, although the theory of addition is decidable, it is actually intractable.

6.6.   SPACE MEASURES AND TRADE-OFFS

Space bounds were already discussed in the previous section. The Turing machines considered in the previous section were deterministic. Space bounds are defined for *nondeterministic* Turing machines in the same way as time bounds.

   The denotations $\mathcal{P}$-SPACE and $\mathcal{NP}$-SPACE stand for languages (or problems) acceptable by deterministic and nondeterministic Turing machines in polynomial space. In Section 6.4 we discussed the celebrated open problem concerning whether or not the corresponding classes of languages with respect to time measure coincide—that is, whether or not $\mathcal{P} = \mathcal{NP}$. For space measure, the same problem is relatively easy to settle:

$$\mathcal{P}\text{-SPACE} = \mathcal{NP}\text{-SPACE}.$$

   We now outline an argument showing how a polynomially space-bounded nondeterministic Turing machine can be simulated by a polynomially space-bounded deterministic Turing machine.

   Let TM be a nondeterministic Turing machine with a polynomial space-bound $P(n)$. There is a constant c (depending on TM) such that whenever TM accepts a word w of length n, the length of the accepting computation does not exceed $c^{P(n)}$. Indeed, c can be chosen as the number of letters of the form $[a, x]$, where a is a tape symbol of TM and x is either a state of TM or 0. Of course, we may restrict the attention to the shortest computation leading to acceptance and, thus, repetitions of the same configuration need not be considered.

   In order to decide deterministically whether or not a specific word is in $L(TM)$, we have to check two words w_0 and w_1 of length $P(n)$ to determine whether or not w_0 yields w_1 according to the rules of TM in less than or equal to $c^{P(n)}$ steps. We have to estimate how much space we need for this checking, whereas time is quite irrelevant.

   A crucial observation is that w_0 yields w_1 in no more than $c^{P(n)}$ steps iff there is a word w_2 of length $P(n)$ such that w_0 yields w_2 in no more than $c^{P(n)}/2$ steps and w_2 yields w_1 in no more than $c^{P(n)}/2$ steps. (For odd numbers c, the smallest integer greater than the actual result of the division might have to be considered.) Similarly, w_0 yields w_2 in no more than $c^{P(n)}/2$ steps iff there is a word w_3 of length $P(n)$ such that w_0 yields w_3 in (essentially) no more than $c^{P(n)}/4$ steps and w_3 yields w_2 in no more than $c^{P(n)}/4$ steps. This procedure is continued until the number of steps is reduced to 1.

How much space does a Turing machine executing this algorithm use? All the intermediate words have to be stored but not permanently: The space can be reused if a particular word has been shown to lead to failure. Thus, at any particular moment, it suffices to store the order of $P(n)$ words of length $P(n)$. (We need here the order of $P(n)$ rather than $P(n)$. Essentially, $P(n)$ has to be multiplied by $\log_2 c$.) Consequently, the space needed is of the order of $(P(n))^2$ and, thus, the space is polynomially bounded.

When we take into account the fact that every function as a time bound yields the same function as a space bound, we obtain the following result.

Theorem 6.23. $\mathcal{P} \subseteq \mathfrak{N}\mathcal{P} \subseteq \mathcal{P}\text{-SPACE} = \mathfrak{N}\mathcal{P}\text{-SPACE}$.

It is a celebrated open problem whether or not the inclusions in Theorem 6.23 are strict.

The notion of a $\mathcal{P}$-SPACE complete language is defined in the same way as an $\mathfrak{N}\mathcal{P}$-complete language. By definition, a language L is $\mathcal{P}$-**SPACE complete** iff L is in $\mathcal{P}$-SPACE and, furthermore, every language L' in $\mathcal{P}$-SPACE is polynomially reducible (with respect to space) to L.

Encode the letter a_i in an alphabet $\{a_1, \ldots, a_r\}$ as $\#i\#$, where i is written as a dyadic integer. Let L_{REG} be the language over the alphabet

$$\{\cup, *, \varnothing, (,), \#, 1, 2\},$$

consisting of all regular expressions (with letters written in the form $\#i\#$) denoting a language L with a nonempty complement (with respect to the minimal alphabet of L).

Theorem 6.24. L_{REG} is $\mathcal{P}$-SPACE complete.

The proof of Theorem 6.24, being almost the same as the proof of Theorem 6.22, is left to the reader. One again constructs a regular expression denoting "noncomputations": If some word fails to be in the set denoted by the regular expression, the word must correspond to an accepting computation. The only new issue is to show that L_{REG} is in $\mathcal{P}$-SPACE. But it is easy to construct a nondeterministic polynomially space-bounded Turing machine accepting L_{REG}. This will be sufficient, by Theorem 6.23.

Comparing Theorems 6.22 and 6.24, we see that adding the operation of squaring to regular expressions is quite essential from the complexity point of view: A problem in $\mathcal{P}$-SPACE (and even in $\mathcal{P}$, provided the inclusions in Theorem 6.23 reduce to equalities) becomes a problem outside $\mathcal{P}$-SPACE.

It is an immediate consequence of Theorem 6.24 that L_{REG} is in $\mathcal{P}$ iff $\mathcal{P} = \mathcal{P}\text{-SPACE}$. It also follows that L_{REG} is in $\mathfrak{N}\mathcal{P}$ iff $\mathfrak{N}\mathcal{P} = \mathcal{P}\text{-SPACE}$.

The theory of **logarithmic space** is a highly developed subarea in studies concerning space bounds. A language is in LOGSPACE (resp. NLOGSPACE) iff it is accepted by a deterministic (resp. nondeterministic) Turing machine whose space complexity function is bounded from above by a function of the order of log n. Since an input of length n takes up n squares by itself, we have to consider here a model of Turing machines, where input and output are given on separate tapes (the former is given on a read-only tape, and the latter is produced on a write-only tape), whereas the actual computation is performed on a work tape. The space complexity refers to the number of squares used on the work tape.

It is not difficult to prove that

$$\text{LOGSPACE} \subseteq \mathcal{P}.$$

It is an open problem whether or not the inclusion is strict. However, it is known that at least one of the inclusions

$$\text{LOGSPACE} \subseteq \mathcal{P} \quad \text{and} \quad \mathcal{P} \subseteq \mathcal{P}\text{-SPACE}$$

is strict—This is a consequence of space hierarchy results (see Exercise 6.16).

We earlier introduced the notion of polynomial reducibility (with regard to time). Similarly, we may introduce **log-space reducibility.** By definition, a language L_1 is log-space reducible to a language L_2—in symbols

$$L_1 \leq_{LS} L_2,$$

—iff there is a log-space-bounded deterministic Turing machine translating words in such a way that the input being in L_1 is equivalent to the output being in L_2. Since log-space computation can also be done in polynomial time,

$$L_1 \leq_{LS} L_2 \quad \text{implies} \quad L_1 \leq_P L_2.$$

It is not difficult to see that the relation $\leq_{LS}$ is transitive and that, whenever $L_1 \leq_{LS} L_2$ and L_2 is in LOGSPACE, then L_1 is also in LOGSPACE.

A language L is termed **log-space complete** for $\mathcal{P}$ iff L is in $\mathcal{P}$ and every L' in $\mathcal{P}$ satisfies $L' \leq_{LS} L$. Thus, $\mathcal{P} = \text{LOGSPACE}$ iff there is a language log-space complete for $\mathcal{P}$ that is actually in LOGSPACE.

An example of a problem log-space complete for $\mathcal{P}$ is the emptiness problem of a context-free language. More explicitly, the language L consisting of (suitably coded versions of) context-free grammars generating the empty language is log-space complete for $\mathcal{P}$.

The notion of log-space completeness for NLOGSPACE is defined analogously. An example of a problem log-space complete for NLOGSPACE is the following graph-reachability problem. Given a finite directed graph and two specific nodes BEGIN and END, determine whether or not there is a path from BEGIN to END.

Clearly, LOGSPACE ⊆ NLOGSPACE, but it is not known whether or not the inclusion is strict. It is known that NLOGSPACE ⊆ $\mathcal{P}$. Also, the following results concerning the comparison between space and time are known.

Denote by POLYLOGSPACE the family of all languages acceptable by a deterministic Turing machine whose space complexity function is bounded from above by the function $\log^k n$, for some $k \geq 1$. Then it can be shown that

$$\mathcal{P} \neq \text{POLYLOGSPACE} \quad \text{and} \quad \mathcal{N}\mathcal{P} \neq \text{POLYLOGSPACE}.$$

However, it is known only that the sets are different; nothing further is known about the mutual relationship between sets.

Finally, we mention some basics about a method for discussing relationships between time and space. Clearly, trade-offs are possible: Allowing more time may result in less space, and vice versa.

The method is referred to as **pebbling**. Basically, it consists of a game, called **pebble game**, played on a graph. In fact, this game can be viewed as a one-person game. The graph involved is a directed one. A node into which no edges are directed is termed an *input*. Similarly, a node from which no edges are emanating is termed an *output*.

At any time instant, some nodes will have pebbles on them. An **instantaneous description** consists of the subset of nodes having pebbles on them.

The pebble game has the following two rules.

Rule 1. A pebble may be removed from any node.

Rule 2. A pebble may be placed on a node a, provided every node x such that there is an arrow from x to a has a pebble on it.

A **computation step** is an ordered pair of instantaneous descriptions such that the second one follows from the first according to our two rules. A **computation** is a finite sequence of computation steps. A **proper computation** is a computation beginning and ending with empty instantaneous descriptions and in which every node appears in some instantaneous description. The **time** in a computation is the number of steps, and the **space** is the maximal number of nodes in any instantaneous description.

A basic result in this context is that a computation in time bound $T(n)$ can be simulated by a computation, where the space bound is of the order $T(n)/\log n$. The reader is referred to [Pi] for various results concerning trade-offs between time and space. The results deal either with special types of graphs, graphs corresponding to well-known algorithms, or, more generally, graphs corresponding to any algorithm that might be devised for solving a given problem.

EXERCISES

6.1.   Show that any effective listing of all Turing machines yields an acceptable enumeration (3). Show that appropriate versions of Theorems 4.1 and 4.3 (*s-m-n* theorem and universal machine theorem) hold for every acceptable enumeration.

6.2.   Does the number of moves to the right during a computation of a Turing machine constitute a complexity measure? What about the number of turns the read-write head makes during a computation?

6.3.   Show that the recursion theorem (Theorem 4.6) holds for every acceptable enumeration. (Actually, the proof of Theorem 4.6 can almost be used as such.)

6.4.   Show that Theorem 6.9 holds with the additional condition that the functions $f_{h(i)}$ assume only the values 0 and 1.

6.5.   Show that the condition "almost everywhere" is unavoidable in Theorem 6.10.

6.6.   Establish a modification of the speedup theorem (Theorem 6.10), where the function $f(x)$ assumes only the values 0 and 1.

6.7.   Apply Theorems 4.14 and 5.9 to show that the enumeration (based on polynomials) discussed before Theorem 6.12 constitutes an acceptable enumeration of all partial recursive functions.

6.8.   Consider algorithms with time complexities n, n^2, n^3, 2^n, and 3^n and problems of sizes $n = 100$, $n = 10^4$, $n = 10^6$, and $n = 10^8$. Write down a table of computing times, assuming that the computer performs one operation per microsecond. Estimate also, for the five time complexities indicated, the increase in the manageable problem size provided the computer becomes ten times as fast.

6.9.   Give explicit time estimates for Theorem 6.15.

6.10. Show that the relation $\leq_P$ is transitive. Show also that

$$L_1 \leq_P L_2 \quad \text{implies} \quad N(L_1) \leq_m N(L_2),$$

where $N(L_i)$ is the set of integers associated to L_i defined in Section 4.4 and $\leq_m$ is the *m*-reducibility defined in Section 4.5. Use intractability results of Section 6.5 to show that the converse implication is not valid.

6.11. Show that the following problems concerning undirected graphs are $\mathfrak{NP}$-complete. Does a given graph have a clique of size k? (A *clique* means a set of nodes in which every two nodes are connected with an edge.) Is a given graph k-colorable? Does a given graph have a Hamilton circuit—that is, a cycle containing every node?

6.12. Prove that the following traveling salesperson problem is $\mathfrak{NP}$-complete: Given a directed graph with integer weights associated to the edges and an integer k, determine whether or not the graph has a Hamilton circuit with weight $\leq k$.

6.13. Show that the following integer linear programming problem is $\mathfrak{N}\mathcal{P}$-complete: Given an $m \times n$ matrix M with integer entries and an integer column vector $\mathbf{v}$ of dimension m, does there exist an integer column vector $\mathbf{x}$ such that $M\mathbf{x} \geq \mathbf{v}$? It is a celebrated recent result of [Kh] that this problem is in $\mathcal{P}$ if $\mathbf{x}$ is required to be a rational vector. (This modification is usually referred to as *ordinary* linear programming.)

6.14. Denote by CO-$\mathfrak{N}\mathcal{P}$ the family of languages L such that the complement $\sim L$ of L is in $\mathfrak{N}\mathcal{P}$. The relation between $\mathfrak{N}\mathcal{P}$ and CO-$\mathfrak{N}\mathcal{P}$ is unknown. Prove that if there exists an $\mathfrak{N}\mathcal{P}$-complete language whose complement is in $\mathfrak{N}\mathcal{P}$, then $\mathfrak{N}\mathcal{P} =$ CO-$\mathfrak{N}\mathcal{P}$.

6.15. Prove that if there is an $\mathfrak{N}\mathcal{P}$-complete language over a one-letter alphabet, then $\mathcal{P} = \mathfrak{N}\mathcal{P}$.

6.16. Prove the existence of a language L such that L is accepted by a deterministic Turing machine with space bound 2^n, but it is accepted by no deterministic Turing machine with space bound $2^{n/2}$. (This is a special case of the space hierarchy theorem for Turing machines. The theorem says, essentially, that a big enough increase in the space bound brings in new languages. The proof is by diagonalization.)

6.17. Prove that if $\mathcal{P} \neq \mathfrak{N}\mathcal{P}$, then $\mathfrak{N}\mathcal{P}$ contains languages that are neither in $\mathcal{P}$ nor $\mathfrak{N}\mathcal{P}$-complete.

6.18. Prove that the membership problem of a D0L language is in LOGSPACE. (The comprehensive area concerning the complexity of the membership problem of languages in a specific language family has not been discussed in this book.)

CHAPTER 7

Cryptography

7.1. BACKGROUND AND CLASSICAL CRYPTOSYSTEMS

It might seem strange that a chapter on cryptography appears in a book dealing with the theory of computation, automata, and formal languages. However, in the last two chapters of this book we want to discuss some recent trends. Undoubtedly, cryptography now constitutes such a major field that it cannot be omitted, especially because its interconnections with some other areas discussed in this book are rather obvious. Basically, cryptography can be viewed as a part of formal language theory, although it must be admitted that the notions and results of traditional language theory have so far found only few applications in cryptography. Complexity theory, on the other hand, is quite essential in cryptography. For instance, a cryptosystem can be viewed as *safe* if the problem of cryptanalysis—that is, the problem of "breaking the code"—is intractable. In particular, the complexity of certain number-theoretic problems has turned out to be a very crucial issue in modern cryptography. And more generally, the seminal idea of modern cryptography, public key cryptosystems, would not have been possible without an understanding of the complexity of problems. On the other hand, cryptography has contributed many fruitful notions and ideas to the development of complexity theory.

During recent years the increase in the number of studies concerning cryptography has been explosive; there are two main reasons for this. In the first place, public key cryptography, apart from being of extraordinary interest on its own right, has created entirely new types of applications, some of which are rather contraintuitive with respect to the common-sense notions concerning communication between different parties. Such applications are discussed later on in this chapter, especially in Section 7.5. In the second place, the need for cryptography has increased tremendously due to various aspects of data security. This can be expressed briefly by saying that there are now really many positive applications of cryptography, whereas the earlier applications were mainly negative—that is, secret communication during wars, criminal activities, and so on.

Let us elaborate the second reason mentioned. The positive applications to which we refer are all somehow connected with the very important task of assuring that information stored in an information system can be accessed by proper, or authorized, persons only.

Two different methods are possible for achieving this goal: access-path control and data encryption. The former method involves preventing unauthorized access by building appropriate controls, such as secret passwords that can (it is to be hoped!) be passed only by authorized users, into the system. There are a number of obvious disadvantages inherent to this approach. For instance, the data processing personnel familiar with the system may be able to circumvent access-path controls. Besides, data stored externally (punched cards, magnetic tapes, and so on) have to be protected as well, and only real physical protection seems to be possible for this purpose. Thus, if an extreme level of security is required (for instance, for some military purposes), then access-path control is apparently possible only in the sense of real physical protection. The latter, on the other hand, is very impractical in connection with large information systems.

However, access to information can also be guarded by the other method mentioned: data encryption. This means that information is fed in the system in an encrypted form. Although the encrypted version of information is obtainable by unauthorized persons, it is useless to anyone not knowing the decryption technique. Hence, assuming that hard-to-break encryption techniques are used, the system will be safe against unauthorized users. Moreover, the method of data encryption has several practical advantages, as will be seen later on.

We are now ready to begin the actual discussion of cryptography. We first introduce some terminology needed throughout this chapter.

Consider messages sent via an insecure channel. (The channel might be a telephone line subject to wiretapping by an eavesdropper, or it might be some data-storage device in an information system.) Such messages will be referred to as **plaintext**. To increase security, the sender **encrypts** the plaintext and sends the resulting **cryptotext** via the channel. The receiver then **decrypts** the cryptotext back to the plaintext. (Instead of plaintext and cryptotext, the terms **cleartext** and **ciphertext** are often used.)

The encryption and decryption are usually carried out in terms of a specific *cryptosystem*. Essentially, a cryptosystem specifies several (often infinitely many) **keys.** Each key K determines an encryption function e_K and a decryption function d_K. Cryptotext c is obtained from plaintext w using e_K:

$$e_K(w) = c.$$

Conversely,

$$d_K(c) = w.$$

Thus,

$$d_K(e_K(w)) = w. \tag{1}$$

In this sense, d_K is an inverse of e_K.

More specifically, a **cryptosystem** CS consists of a **plaintext-space,** a **cryptotext-space,** and a **key-space.** All three items are at most denumerable. Typically, a plaintext-space could be Σ^*, for some alphabet Σ, or the set of all meaningful sentences in a natural language, such as English. Similarly, the cryptotext-space could be Δ^*, for some alphabet Δ. Each key K determines mappings e_K and d_K in the sense discussed above. For instance, if the plaintext- and cryptotext-spaces are Σ^* and Δ^*, then e_K will be a translation of Σ^* into Δ^*.

It should be obvious to the reader that the preceding paragraph is intended to be an intuitive description rather than a formal definition of a cryptosystem. None of the three items has been defined formally, and neither has the way in which the mappings e_K and d_K are associated with keys K. Indeed, in our estimation the state of art with regard to foundations of cryptography has not yet reached a level where most of the phenomena in interesting practical examples can be described in a satisfactory fashion. In particular, it seems that the interrelation between complexity-theoretic and information-theoretic aspects is not yet clearly understood. This point will also be discussed in the sequel.

Consequently, we shall be satisfied with an informal notion of a cryptosystem, also illustrated by several examples. In fact, we shall allow so much leeway that it is not even necessary for e_K and d_K to be functions. This is due to the fact that encryption in many ''good'' classical cryptosystems is not monogenic; the same plaintext can be encrypted in several ways by the same key. However, in most cases e_K and d_K will be functions satisfying (1).

We next discuss some intuitively clear grounds that make a cryptosystem good or bad. We refer to some problems as *easy, hard,* or *intractable.* This is to be understood in the sense of the complexity considerations of the previous chapter. From the point of view of cryptography, an $\mathcal{NP}$-complete problem would be considered as intractable, whereas a problem being easy would require some low polynomial bound.

Given the key K and the plaintext w, the computation of the crypto-text $c = e_K(w)$ should be easy in a good cryptosystem. So should be the computation of the plaintext w from the equation $w = d_K(c)$. This means that the legal receiver should be able to recover the plaintext from the cryptotext without too much trouble. From the point of view of storing and sending messages, another reasonable requirement for a good cryptosystem is that the cryptotext $e_K(w) = c$ should not be very much longer than the plaintext w. Some authors even demand that c and w should be of the same length, a rather unreasonable requirement in view of many widely used cryptosystems.

We now come to the essential condition for a cryptosystem to be good: It should be hard, preferably intractable, to recover the plaintext w from the cryptotext c without knowing d_K. The task of recovering w from c is referred to as **cryptanalysis**. (Thus, a cryptanalyst can be viewed as the eavesdropper of our previous discussion.)

The difficulty of cryptanalysis might depend essentially on the preconditions. In what follows, we mention *five possible initial setups* for a cryptanalyst.

(i) *Cryptotext only.* Here the cryptanalysis has to be based on only one sample of cryptotext—that is, the plaintext w has to be recovered from the cryptotext c alone. However, in this as well as in the following approaches, it is customarily assumed that the cryptanalyst knows the cryptosystem used. This assumption is reasonable in view of the fact that, for instance, when users are storing data in encrypted form in an information system, a specific cryptosystem apparently has to be applied.

(ii) *Known plaintext.* Here the cryptanalyst knows in advance some pairs $(w, e_K(w))$. The knowledge of such pairs may essentially aid the analysis of the given cryptotext c.

(iii) *Chosen plaintext.* The cryptanalyst knows in advance some pairs $(w, e_K(w))$, but now in these pairs the plaintext w is not arbitrary: It has been chosen by the cryptanalyst. In situations where the cryptanalyst has definite conjectures about the key, it is clear that this setup is essentially better than (ii). On the other hand, this setup (iii) is likely to be realistic at least in such cases where the cryptanalyst has the possibility of masquerading himself or herself as an authorized user of the information system in question.

The last two setups, (iv) and (v), refer to public key cryptosystems only. Public key systems are discussed in much more detail below. Here, we mention only their characteristic feature: Even if the encryption function e_K is known, it is still intractable to find the decryption function d_K.

(iv) *Chosen cryptotext.* The cryptanalyst knows in advance some pairs $(c, d_K(c))$, where the cryptotext c has been chosen by the cryptanalyst. (Of course, some choices of c may lead to a meaningless plaintext $d_K(c)$.)

(v) *Encryption key only.* The cryptanalyst knows the encryption function e_K and tries to find d_K before actually receiving any samples of cryptotext. The analyst might have plenty of time for doing this because, in the normal case of a public key cryptosystem, e_K is made public in advance, and it might take a long time before e_K is used to encrypt important messages. Thus, in this setup, the cryptanalyst might break the cryptosystem by **preprocessing**—that is, before receiving the encrypted message in which he or she is actually interested. Of course, preprocessing might reveal the decryption function d_K also in setups (ii)–(iv).

As we already pointed out, in all the setups (i)–(v), it is customarily presupposed that the cryptanalyst knows what cryptosystem has been used. Indeed, this supposition is reasonable because the cryptosystem is usually known in connection with data encryption and, moreover, the complexity of cryptanalysis is increased by only a constant factor even if a fixed number of cryptosystems has to be tried.

The notion of a cryptosystem is now illustrated by several examples.

Example 7.1. We begin with perhaps the oldest and best-known among all cryptosystems, **Caesar cipher** or, briefly, CAESAR. (In the sequel we shall often write the names of cryptosystems in capital letters.)

In fact, CAESAR was already discussed in Example 2.3. In this case, both the plaintext-space and the cryptotext-space consist of all words over the English alphabet $\Sigma = \{A, B, \ldots, Z\}$. The key space is $\{0, 1, 2, \ldots, 25\}$. If K is an element of the key space, then e_K is the morphism mapping each letter K letters ahead in the alphabet. (The ordering of Σ is the natural alphabetic one, and the end of the alphabet is continued cyclically to the beginning.) For instance,

$$e_{25}(\text{IBM}) = \text{HAL}, \qquad e_6(\text{MUPID}) = \text{SAVOJ},$$
$$e_3(\text{HELP}) = \text{KHOS}, \qquad e_0(\text{CRYPTO}) = \text{CRYPTO}.$$

(In Example 2.3, the notation h_{25} was used instead of e_{25}.) The decryption function d_K corresponding to the key K is defined by

$$d_K = e_{26-K}.$$

Thus,

$$d_6(\text{SAVOJ}) = e_{20}(\text{SAVOJ}) = \text{MUPID}.$$

In this case all functions e_K and d_K are clearly bijections of Σ^* onto Σ^*.

The cardinality of the key space of CAESAR is very small. No cryptosystem with this property can be good in the sense described above: Cryptanalysis can be carried out simply by trying all possible keys. This is particularly easy if the plaintext is in some natural language. Then, for reasonably long cryptotexts (say with length greater than or equal to 10), in general only one decryption key produces a meaningful plaintext. This redundancy of natural languages constitutes the foundation for the cryptanalysis of many cryptosystems; knowledge of the statistical distribution of individual letters, pairs of consecutive letters, and triples of consecutive letters uncovers the correspondence between plaintext and cryptotext letters. Details of this technique can be found, for instance, in [Ko].

Of the numerous modifications of CAESAR, we mention the system KEYWORD-CAESAR. Here the key is some previously chosen word, say the word *PLAIN*. This word determines a sequence of integers (15, 11, 0, 8, 13) according to the positions of the letters in the alphabet. We consider A to be the 0th letter—then P is the 15th letter, L the 11th letter, and so on. The plaintext is now encrypted by using the morphisms

$$e_{15}, e_{11}, e_0, e_8, e_{13}, e_{15}, e_{11}, e_0, e_8, e_{13}, e_{15}, \ldots$$

in this order. Thus, the plaintext SENDMOREMONEY becomes the cryptotext HPNLZDCEUBCPY. The Vigenère table given in Example 2.3 can be used in an obvious fashion for encryption and decryption, provided, of course, that the keyword is known. The cardinality of the key space (the number of keywords) is potentially infinite for the system KEYWORD-CAESAR.

Example 7.2. In the cryptosystem SQUARE, keywords are used in an entirely different fashion. We use the English alphabet without the letter Q. (Q is omitted because we want the total number of letters to be a square.) It is now required that all letters in the keyword are different. Our previous keyword PLAIN satisfies this requirement. All letters of the alphabet are now written in the form of a square, beginning with the keyword and followed by the remaining letters in alphabetic order:

P	L	A	I	N
B	C	D	E	F
G	H	J	K	M
O	R	S	T	U
V	W	X	Y	Z

Each pair of plaintext letters is encrypted using this square. For instance, the plaintext SYMPHONY is encrypted as the cryptotext TXGNGRIZ; that is, for each pair of letters, say *NY*, you take the remaining corner letters in the rectangle determined by *NY*: *IZ*. This is the basic idea behind the cryptosys-

tem SQUARE. The reader is encouraged to present a formally detailed encryption technique for this cryptosystem. Essentially, one has to give a deterministic procedure of encryption such that decryption will be unique. Pairs
of letters (α, β) will be encrypted using the square above. If α and β lie in both
a different row and a column, then the encryption of the pair is obvious: Just
take the remaining corners of the rectangle determined by α and β, in the
proper order. If α and β lie in the same row or in the same column, then take,
cyclicly, the next letters in the same row or column. A special convention,
such as adding a dummy letter, has to be applied to prevent the situation
$\alpha = \beta$.

Example 7.3. We now consider a cryptosystem based on linear algebra, due
to Hill. The plaintext- and cryptotext-spaces are again both equal to Σ^*,
where Σ is the English alphabet. We again number the letters in the alphabetic order: A gets the number 0 and Z the number 25.

All arithmetic operations are carried out modulo 26, or modulo the
total number of letters. We first choose an integer $d \geq 2$. It indicates the dimension of the matrices involved. In the encryption procedure, d-tuples of
letters of the plaintext are encrypted together. In what follows d will be 2.

We now choose a d-dimensional square matrix M with integer entries
between 0 and 25 such that M^{-1} (mod 26) exists. For instance,

$$M = \begin{pmatrix} 3 & 3 \\ 2 & 5 \end{pmatrix},$$

implying

$$M^{-1} = \begin{pmatrix} 15 & 17 \\ 20 & 9 \end{pmatrix}.$$

The encryption is now carried out by the equation

$$MP = C, \tag{2}$$

where $\mathbf{P}$ and $\mathbf{C}$ are d-dimensional column vectors. More specifically, each
d-tuple of plaintext letters defines the vector $\mathbf{P}$ where the components are the
numbers of the letters. Then the vector C is computed from (2). Finally, $\mathbf{C}$ is
interpreted as a d-tuple of cryptotext letters.

For instance, the plaintext HELP defines the two vectors

$$\mathbf{P}_1 = \begin{pmatrix} 7 \\ 4 \end{pmatrix} \quad \text{and} \quad \mathbf{P}_2 = \begin{pmatrix} 11 \\ 15 \end{pmatrix}.$$

From the equations

$$M\mathbf{P}_1 = \begin{pmatrix} 7 \\ 8 \end{pmatrix} = \mathbf{C}_1 \quad \text{and} \quad M\mathbf{P}_2 = \begin{pmatrix} 0 \\ 19 \end{pmatrix} = \mathbf{C}_2$$

we obtain the cryptotext HIAT. (Recall that the operations are carried out modulo 26.) The plaintext can be recovered by the equations

$$M^{-1}\mathbf{C}_1 = \mathbf{P}_1 \quad \text{and} \quad M^{-1}\mathbf{C}_2 = \mathbf{P}_2.$$

Indeed, it is immediately verified that

$$\begin{pmatrix} 15 & 17 \\ 20 & 9 \end{pmatrix}\begin{pmatrix} 7 \\ 8 \end{pmatrix} = \begin{pmatrix} 7 \\ 4 \end{pmatrix} \quad \text{and} \quad \begin{pmatrix} 15 & 17 \\ 20 & 9 \end{pmatrix}\begin{pmatrix} 0 \\ 19 \end{pmatrix} = \begin{pmatrix} 11 \\ 15 \end{pmatrix}.$$

Hill's system is a typical example of a cryptosystem, where a particular setup in the cryptanalysis makes the system easy to break. Here the setup is chosen plaintext. The cryptanalyst chooses d plaintext vectors (for $d = 2$, the vectors $\mathbf{P}'$ and $\mathbf{P}''$) such that the square matrix M' formed by these vectors possesses an inverse. Let C' be the square matrix formed by the cryptotext vectors corresponding to the chosen plaintext vectors. The key M can now be immediately computed from the equation

$$MM' = C'. \tag{3}$$

From the encryption key M, the decryption key M^{-1} is also immediately obtainable. (Observe that in cryptography one always has to be aware of the complexity of the computations involved. The computations in linear algebra resulting from (3) can be done very fast.)

The setup known plaintext also makes Hill's system easy to break. If somewhat more than d plaintext vectors are known together with the corresponding cryptotext vectors, then some of the resulting matrices M' possess an inverse with a high probability. On the other hand, the cryptanalysis is not so obvious if the setup is cryptotext only.

Example 7.4. In the following cryptosystem, HOMOPHONIC-CRYPTO, the encryption procedures e_K are not actually functions. However, the system has the important advantage that the obvious cryptanalytic approach, based on the statistical distribution of letters in the plaintext language, is not possible as such.

The plaintext- and cryptotext-spaces are the same as in Example 7.3. With each plaintext letter α, we associate some numbers, called **homophones** of α, between 0 and 99. The encryption procedure is now obvious: Each letter α is replaced by some number i, picked at random from the set of homophones of α. The point in this procedure is that, for each letter α, the number of homophones assigned to α is proportional to the relative frequency of α. (Of course, no homophone is assigned to more than one letter.)

For instance, the assignment might be as in Table 7.1. We list only the homophones for the letters actually appearing in the plaintext considered on page 194.

TABLE 7.1

Letter	Homophones
A	18, 19, 37, 42, 55, 61, 63, 87
I	05, 23, 28, 73, 75, 91
L	03, 43, 92
N	11, 15, 31, 49, 56, 66, 95
O	04, 12, 34, 52, 65, 82, 84, 94
P	33, 98
S	02, 09, 17, 22, 38, 59, 79, 86
T	01, 13, 25, 26, 39, 47, 62, 77, 97

Thus, the plaintext ALLTOPLIONS can be encrypted as

$$1943922682984373841122.$$

We mention, finally, that although the system HOMOPHONIC-CRYPTO is not likely to be broken by a simple frequency analysis based on the distribution of individual letters, a cryptanalysis based on the distribution of pairs of letters, triples of letters, and so on, might still be successful.

Example 7.5. In this final example, we want to emphasize a very important theoretical point. We do not want to discuss the information-theoretic background of cryptography in this book. The core result in information theory is that, with every practically feasible cryptosystem, one can associate a positive integer n, referred to as the **unicity distance** of the cryptosystem, such that every cryptotext with length at least n uniquely determines the key in use. This result is based on the assumption that the plaintext is in a language with some redundancy. Of course, every natural language has this property.

The unicity distance can be approximated by the ratio of the a priori uncertainty in the key to the redundancy (per letter) of the plaintext language.

Consequently, every cryptosystem is rendered insecure if it is used to encrypt too-long plaintexts (too long compared with the unicity distance). It should be emphasized, however, that this argument is merely information-theoretic and does not take into account any considerations based on complexity theory: The breaking of the system might still be computationally quite infeasible. On the other hand, the complexity of cryptanalysis is clearly the main factor making a cryptosystem secure. The relationship between the cryptographic security and computational complexity of cryptanalysis is not yet well understood; the interaction between the information-theoretic and complexity-theoretic aspects involved seems to be hard to formalize.

If variable-length keys are used, then a situation may be reached where the cryptotext gives away no information about the plaintext. The following is an extreme example. Assume that the plaintext is a sequence w of n

bits. The key K is a random sequence of n bits. For every i, the ith bit in the cryptotext is obtained by summing up the ith bits of w and K. (By definition, $1 + 1 = 0$.) The key K is used for the encryption of one plaintext only. This method (called **one-time pad**) has the obvious disadvantage that the key (equally long as the plaintext!) has to be communicated to the receiver via some secure channel.

The following system due to Hammer [Ham], related to HOMOPHONIC-CRYPTO, has the interesting property that arbitrarily long cryptotexts can always be decrypted as *two* meaningful plaintexts. The key is a 26×26 matrix with distinct three-digit entries and with rows and columns indexed by the letters of the English alphabet, for instance

	A	B	C	D ...
A	342	681	082	502
B	431	111	637	249 ...
C	003	228	446	321
D	674	121	913	585
⋮				

The numbers appearing in the row indexed by a particular letter are viewed as homophones of that letter. However, in the encryption procedure, the homophones are not chosen at random, but by applying the following method. For a plaintext (say ADD) another meaningful text of the same length is chosen (say CAB). For encryption, the letters of the plaintext are read from the rows, and the letters of the other meaningful text from the columns. For our example, we get the cryptotext

$$082674121.$$

The decryption is obvious, while the cryptanalyst either misses one possible plaintext or else cannot make, at least not in general, a choice between two plaintexts. □

These examples are by no means intended to give an exhaustive exposition about the different types of classical cryptosystems. They merely serve the purpose of giving an intuitive picture of the most typical phenomena occurring in classical cryptosystems. In particular, we have not discussed at all the system of **data encryption standard,** DES, which is pehaps the most widely used and studied cryptosystem. The system is obtained as a complicated composition of substitutions and permutations. DES has been implemented both in software and in hardware. Hardware implementations achieve a very fast encryption rate (millions of bits per second). The reader is referred to [De], where the extensive literature concerning the cryptanalysis of DES is also quoted.

We mention, finally, some of the possible classifications of cryptosystems. A cryptosystem is referred to as **context-free** or **context-sensitive,** depending on whether or not the encryption of every single plaintext letter is independent of the adjacent plaintext letters. It is **monoalphabetic** or **polyalphabetic,** depending on whether or not two occurrences of the same (sufficiently long and suitably situated) plaintext portion always yield the same cryptotext portion.

CAESAR is an example of a context-free, monoalphabetic system, whereas KEYWORD-CAESAR is an example of a context-free, polyalphabetic system. Hill's system and SQUARE are examples of context-sensitive, monoalphabetic systems. An obvious modification of SQUARE, where different squares are used in some previously agreed periodic sequence, is a context-sensitive, polyalphabetic system.

7.2. PUBLIC KEY CRYPTOSYSTEMS

A property shared by all classical cryptosystems is that whenever you know the encryption key, e_K, then you also know the decryption key, d_K. More explicitly, d_K is either immediately implied by e_K or else can be computed from e_K without too much difficulty.

For instance, in the cryptosystems CAESAR and KEYWORD-CAESAR, as well as in the cryptosystem SQUARE, the encryption key immediately gives away the decryption key. In regard to Hill's system (Example 7.3), you have to compute the inverse of a given matrix to obtain d_K.

Thus, if you are dealing with any of the classical cryptosystems, you must keep your encryption key e_K secret: Once you publicize e_K, you also give away the decryption key, d_K. We may agree that *classical* in connection with cryptosystems means up to 1975.

The idea of public key cryptosystems (as opposed to classical systems) is due to Diffie and Hellman: The knowledge of e_K does not necessarily give away d_K. More specifically, the cryptanalyst may know e_K and, hence, in principle also the inverse of $e_K(d_K)$. However, the computation of d_K from e_K may be intractable, at least for almost all keys K. Hence, for all practical purposes, the encryption of the plaintext remains secure against cryptanalytic attacks, although the encryption key e_K is made public.

Functions e_K having the property described (that is, the computation of the inverse of e_K is intractable even if e_K is known) are called **one-way** functions. If e_K is a one-way function then the pair (e_K, d_K) is called a **DH-pair** (DH refers to Diffie and Hellman). This terminology is extended to the case where e_K is not necessarily a function; that is, the terminology is extended to concern systems such as HOMOPHONIC-CRYPTO presented in Example 7.4. (As pointed out earlier, we do not necessarily require in the notion of a cryptosystem that the encryption technique e_K is actually a function, although this is the case for most cryptosystems.)

The basic idea behind a public key cryptosystem is, thus, amazingly simple. On the other hand, no reasonable realizations of the idea were conceivable before an adequate development of the theory of computational complexity.

Before discussing the existence of one-way functions and DH-pairs—that is, before trying to give examples of them—we want to point out in general terms the obvious advantages of the resulting public key cryptography.

Consider an information system (or a data bank) with many users A, B, C, ... Each user publicizes, say in a newspaper, his or her encryption key e_A, e_B, e_C, . . ., but keeps the decryption key secret. If some user wants to send a message w to A, he or she feeds it in the system in the form $e_A(w)$. Only A, who knows the secret decryption key d_A, can recover the message: $w = d_A(e_A(w))$.

Compare this situation with the one occurring if a classical cryptosystem is used. In the latter case, the two parties involved must agree in advance upon the secret key that has to be transferred via some secure channel—for example, during an a priori meeting of the parties. In case of a public key cryptosystem, no previous meeting of the parties is needed. In fact, it is not necessary for the parties or persons involved even to know each other! The whole action can be based on a newspaper announcement. This is a tremendous advantage for any information system with many users with different types of authorization, such as the information system handling the activities of a big bank.

A nice mechanical analog to depict the difference between classical and public key cryptosystems is the following. Assume the information is sent in a box with clasp rings. Then, encryption according to a classical cryptosystem corresponds to the locking of the box with a padlock and sending the key via some absolutely secure channel, such as using an agent like James Bond. Public key cryptography corresponds to having open padlocks, provided with your name, available in places such as post offices. A person who wants to send a message to you can close a box with your padlock and send it to you. Only you have a key for opening the padlock.

An interesting modification of the basic public key procedure is the following. Assume that the cryptosystem is **commutative** in the sense that in any composition of the functions e_A, e_B, . . ., as well as their inverses, the order of the factors is immaterial. Suppose A wants to send the message w to B. This can happen according to the following protocol:

    (i) A sends $e_A(w)$ to B.
    (ii) B sends $e_B(e_A(w))$ to A.

(iii) A sends $d_A(e_B(e_A(w))) = d_A(e_A(e_B(w))) = e_B(w)$ to B.

(iv) B decrypts: $d_B(e_B(w)) = w$.

Observe that this protocol is, in fact, suitable for classical cryptosystems as well: Only commutativity is required. Thus, open padlocks need not be distributed in advance. First, A sends the box to B, locked with the padlock e_A. Then, B sends the box back to A, also locked with e_B. Next, A opens e_A and sends the box back to B. Now, B can open it. Thus, the box is always protected by at least one padlock. (This analogy is due to Michael Rabin.)

The protocol described above is secure against **passive** eavesdroppers. However, an **active** eavesdropper C might masquerade himself or herself as B. Then A has no way of knowing who the other party actually is. (By a *passive eavesdropper* we mean a cryptanalyst who tries only to obtain all possible information in order to decipher important messages. An *active eavesdropper* masquerades himself or herself as the intended receiver of a message and returns information to the original sender accordingly.)

Thus, to avoid the threat from active eavesdroppers, some authentication procedure is needed. There are also many similar situations, where a message planted in a communication channel or an information system should be somehow "signed." If electronic mail systems are going to replace the existing paper mail system, the need for such **electronic signatures** will be really great. We now discuss in more detail some obvious requirements for electronic signatures, as well as the possibilities of fulfilling these requirements by the use of public key cryptosystems.

Consider two parties A and B with conflicting interests—for example, a bank and a customer or two superpowers. When A sends a message to B, it should be signed and the signature should give the parties the following two kinds of protection.

(i) Both A and B should be protected against messages addressed to B but fed in the information system by a third party C, who pretends to be A.

(ii) A should be protected against messages forged by B, who claims to have received them from A, properly signed.

If a classical cryptosystem is used, then requirement (i) can be satisfied in a reasonable fashion: A and B agree upon a secret encryption key known only to them. On the other hand, requirement (ii) is apparently more difficult to satisfy because B should know something about the way A generates the signature, and yet it should be impossible for B to generate A's signature. Observe also that if we are dealing with a big network of communicating parties (such as a network of mail users) then it is impractical to use a distinct secret method of signing for every pair of users.

If a public key cryptosystem is used, then both (i) and (ii) can be satisfied, at least in principle. First, A sends the message w to B in the form

$e_B(d_A(w))$. Then, B can recover $d_A(w)$ by his or her secret decryption key d_B. From $d_A(w)$, B can recover w by the publicly known e_A. (Suppose that e_A and d_A are inverses.) Now both (i) and (ii) are satisfied because only A knows d_A. Hence, neither C nor B can forge A's signature. Also, A cannot later deny sending the message to B.

If only signature (but not encryption) is needed, then the procedure described above can be modified as follows. First, A sends B the pair $(w, d_A(w))$. Then (i) and (ii) are satisfied as before. Of course, in this case even a passive eavesdropper finds out the contents of the message.

Our preceding discussion about the obvious advantages of public key cryptosystems has been on an abstract level because we have not given any examples of public key cryptosystems—we have only stated that it is conceivable that they exist. In what follows, we discuss specific public key cryptosystems.

This discussion begins with a general method of constructing one-way functions. We then present two preliminary examples. Two widely studied public key cryptosystems are presented in Sections 7.3 and 7.4.

The construction of a public key cryptosystem is based on the idea of a **trapdoor:** The information publicized is not enough without the knowledge of a secret trapdoor.

More specifically, the trapdoor technique works as follows in the construction of public key cryptosystems.

(i) Start with a difficult (intractable) problem Q.
(ii) Consider an easy subproblem Q_1 of Q. Q_1 should be in polynomial time, preferably in linear time.
(iii) "Shuffle" Q_1 in such a way that the resulting problem $\overline{Q_1}$ "looks like" the original problem Q. Use $\overline{Q_1}$ as a public encryption key.
(iv) Keep secret the information concerning how Q_1 can be recovered from $\overline{Q_1}$. This information is referred to as the **trapdoor.**

Of course, this technique is still on the abstract level: in practice, the details have to be filled in. However, the public key cryptosystems existing so far have been constructed by this technique.

It is rather amazing that in all examples existing so far, the initial problem Q is at most $\mathfrak{NP}$-complete. (Here "at most" means that the problem is in $\mathfrak{NP}$.) However, it is conceivable—and even desirable—that the initial problem could be much higher up in the complexity hierarchy, perhaps even an undecidable problem. That no examples of such a situation are known is due to the fact that no reasonable encryption techniques corresponding to (iii) have been found for undecidable initial problems.

For instance, Theorem 5.6 shows that the equivalence problem for $0L$ systems is undecidable. On the other hand, the subproblem concerning the equivalence of $D0L$ systems is very easy. In fact, it is generally conjectured that it suffices to test the equality of the languages obtained from the corresponding $D0L$ sequences up to $2n$ words, where n is the cardinality of the alphabet. Consequently, we could choose Q as the equivalence problem for $0L$ systems and Q_1 as the equivalence problem for $D0L$ systems. The difficulty here is that no way of satisfying (iii) is known so far in this case: how to shuffle Q_1 in such a way that the resulting $\overline{Q}_1$ can still be used as a reasonable encryption key.

Observe that $\overline{Q}_1$ only looks like Q but cannot be the same problem, provided Q is not in $\mathcal{P}$. This follows because, clearly, $\overline{Q}_1$ is in $\mathcal{P}$. But the trapdoor information is needed in order to find the polynomial-time algorithm for $\overline{Q}_1$.

Sections 7.3 and 7.4 discuss the most widely studied public key cryptosystems. Both are based on the trapdoor idea.

This section contains two examples and a theoretical result. Example 7.6 is an intuitive example about public key cryptosystems. Example 7.7 concerns an application of formal language theory to public key cryptography.

Example 7.6. The encryption technique considered in this and the next example is similar to the one considered in Example 7.4: Both are nondeterministic in the same sense.

The encryption is based on the telephone directory of a big city. For each letter of the plaintext, a name beginning with that letter is chosen at random from the directory. The corresponding telephone number constitutes the encryption of (the occurrence of) the letter in question.

For instance, the encryption of the plaintext TOKYO is obtained by writing one after the other all numbers appearing in the right column of Table 7.2.

The encryption procedure may be publicized. Cryptanalysis will still be time-consuming because it has to be based, essentially, on an exhaustive search. (Of course, you might also try to call the numbers in the cryptotext and ask the names!) On the other hand, decryption is easy if the trapdoor information, consisting of a directory listed according to the increasing order of the numbers, is available.

TABLE 7.2

Plaintext	Name Chosen	Cryptotext
T	THOMPSON	7184142
O	OWENS	3529517
K	KLEIN	9372712
Y	YYTERI	2645611
O	ORWELL	2193881

Example 7.7. We now consider a public key cryptosystem based on formal language theory—more specifically, on the theory of L systems. Basically, the public encryption key will be a $T0L$ system obtained from an underlying, much-simpler deterministic system—that is, $DT0L$ system. The trapdoor information kept secret concerns how the $DT0L$ system can be recovered from the $T0L$ system. Without this information, cryptanalysis leads essentially to an $\mathfrak{NP}$-complete problem, namely, the membership problem for $T0L$ systems.

The reader interested in the background concerning $T0L$ and $DT0L$ systems is referred to [RozS]. However, the background is not necessary for understanding Example 7.7 because the example will be stated in general terms.

Let Σ be an alphabet and consider two morphisms $h_0, h_1 : \Sigma^* \to \Sigma^*$, as well as a nonempty word w in Σ^*. For a word $i_0 i_1 \cdots i_n$ over the alphabet $\{0, 1\}$, where each i is a letter, and a word x over Σ we denote briefly

$$(x)i_0 i_1 \cdots i_n = h_{i_n}(\ldots h_{i_1}(h_{i_0}(x))\ldots).$$

We assume that our morphisms are unambiguous in the following sense. Whenever

$$(w)i_0 i_1 \cdots i_n = (w)j_0 j_1 \cdots j_m$$

then

$$i_0 i_1 \cdots i_n = j_0 j_1 \cdots j_m.$$

Let Δ be an alphabet of a much-greater cardinality than Σ. (Typically, Σ could consist of two letters, whereas Δ could consist of 200 letters.) Let $g : \Delta^* \to \Sigma^*$ be a morphism mapping every letter to a letter or to the empty word in such a way that $g^{-1}(a)$ is nonempty for all letters a in Σ. Define two substitutions σ_0 and σ_1 on Δ^* by letting $\sigma_i(d)$ to be a finite nonempty subset of $g^{-1}(h_i(g(d)))$ for all d in Δ and $i = 0, 1$.

Intuitively, the morphism g divides the letters of Δ into *descendants* of letters of Σ and into *junk* symbols that are mapped into the empty word.

Finally, choose a word u from $g^{-1}(w)$. The quadruple $(\Delta, \sigma_0, \sigma_1, u)$ now constitutes the **public encryption key**. The encryption of a plaintext $i_0 i_1 \cdots i_n$ (in bits) is carried out by choosing an arbitrary word from

$$\sigma_{i_n}(\ldots \sigma_{i_1}(\sigma_{i_0}(u))\ldots). \tag{4}$$

The plaintext can also be divided into blocks (not necessarily of equal length) that will be encrypted separately by the method described. This latter procedure is recommended if the cryptotext tends to be too long.

The **secret decryption key** consists of the items Σ, h_0, h_1, w, and g. In fact, g is the essential item because the other items can be determined from g and the public encryption key.

It follows from the definition of the public encryption key that whenever x is a word in the set of (4), then

$$g(x) = (w)i_0i_1\cdots i_n.$$

The original plaintext can now be recovered from $g(x)$ using the items Σ, h_0, h_1, and w. (In the terminology of L systems, we make use of the $DT0L$ system determined by these four items. Details are discussed later.)

We now consider a specific example. The morphisms and substitutions are defined in terms of productions. This notation is used to aid intuition.

Assume that $\Sigma = \{a, b\}$, and that h_0 and h_1 are defined by

$$h_0 : a \rightarrow ab, b \rightarrow b; \qquad h_1 : a \rightarrow a, b \rightarrow ba.$$

Choose $w = ab$.

Consider $\Delta = \{c_1, c_2, c_3, c_4, c_5\}$. The alphabet Δ is much too small for any practical purposes. However, practically feasible alphabets would result in systems unreadable in a book. Also, later on in our examples about public key cryptosystems, the reader should be aware of the fact that the examples merely serve the purpose of illustrating a point and are not intended to be public key cryptosystems really applicable in practice.

Define

$$g(c_2) = g(c_4) = a, \qquad g(c_1) = b, \qquad g(c_3) = g(c_5) = \lambda.$$

Thus, c_2 and c_4 are descendants of a, c_1 is the only descendant of b, and c_3 and c_5 are junk letters.

Define σ_0 and σ_1 by

$$\sigma_0: \quad c_1 \rightarrow c_5c_1, \; c_2 \rightarrow c_2c_3c_1, \; c_2 \rightarrow c_4c_5c_1, \; c_3 \rightarrow c_3c_5,$$
$$c_3 \rightarrow c_5c_3, \; c_4 \rightarrow c_2c_1c_3c_5, \; c_5 \rightarrow c_5c_5c_5;$$
$$\sigma_1: \quad c_1 \rightarrow c_1c_2, \; c_1 \rightarrow c_1c_4, \; c_2 \rightarrow c_4, \; c_3 \rightarrow c_3c_5c_3,$$
$$c_3 \rightarrow c_5c_5, \; c_4 \rightarrow c_2, \; c_4 \rightarrow c_4, \; c_5 \rightarrow c_5.$$

Finally, define $u = c_2c_3c_1$. The quadruple $(\Delta, \sigma_0, \sigma_1, u)$ can now be publicized as the encryption key.

Consider the plaintext 010. Thus, the encryption must start from u and use the substitutions σ_0, σ_1, and σ_0. One possible derivation goes as follows:

$$u = c_2c_3c_1 \underset{\sigma_0}{\Rightarrow} c_2c_3c_1c_3c_5c_5c_1$$
$$\underset{\sigma_1}{\Rightarrow} c_4c_5c_5c_1c_2c_3c_5c_3c_5c_5c_1c_2$$
$$\underset{\sigma_0}{\Rightarrow} c_2c_1c_3c_5c_5{}^3c_5{}^3c_5c_1c_2c_3c_1c_3c_5c_5{}^3c_5c_3c_5{}^3c_5{}^3c_5c_1c_4c_5c_1$$
$$= c_2c_1c_3c_5{}^8c_1c_2c_3c_1c_3c_5{}^5c_3c_5{}^7c_1c_4c_5c_1 = x.$$

Thus, the word x constitutes an encryption of the plaintext 010. Clearly, x is not the only possible encryption, since there are some alternatives when applying the substitutions σ_i. For instance, the second word might be $c_4 c_5 c_1 c_5 c_3 c_5 c_1$.

A cryptanalyst who has to find the plaintext 010 when knowing x and the public encryption key might, of course, succeed for such a short plaintext and small alphabet, Δ. However, the difficulties of cryptanalysis should be obvious even in this simple case, especially if cryptanalysis is compared with decryption based on the trapdoor information.

Let us now consider decryption based on the trapdoor. We thus know a word y over the alphabet $\Sigma = \{a, b\}$, where y is obtained by applying the morphism g to the cryptotext. We also know that y is of the form

$$y = (w)i_0 i_1 \cdots i_n. \tag{5}$$

On the basis of this knowledge, we have to recover the plaintext $i_0 i_1 \cdots i_n$.

Observe first that whenever we know h_i ($i = 0$ or $i = 1$) and $w' = h_i(w'')$, then w'' is unique and can immediately be computed. This follows by the definition of h_0 and h_1: "parsing" is unique and can be done quickly. Observe, secondly, that if $w' = h_i(w'')$ and w' contains the subword a^2 (resp. b^2) then necessarily $i = 1$ (resp. $i = 0$). By these observations, we immediately find the value of i_n in (5), as well as a word y' such that

$$y' = (w)i_1 i_1 \cdots i_{n-1},$$

after which the procedure is repeated. (The case where neither a^2 nor b^2 is a subword of some of the words y considered is easy to handle: In such a case, the word y must be of the form $ababa \cdots ba$, and the morphism must be h_1.)

We now return to the cryptotext x derived above. We use the trapdoor information. We first obtain

$$y = g(x) = ab^2 ab^2 ab.$$

Since b^2 occurs as a subword in y, we conclude that 0 is the last letter of the plaintext. The word y' satisfying $y = h_0(y')$ is $y' = ababa$. This reveals 1 to be the following plaintext letter, counting from the end. Again, the corresponding y'' is found:

$$y'' = ab^2 \quad \text{satisfies} \quad y' = h_1(y'').$$

The subword b^2 in y'' tells us that the next plaintext letter is again 0. Since $h_0(ab) = y''$, we have finished the decryption, obtaining the plaintext 010.

The analysis given above also shows that the morphisms h_0 and h_1 considered in our specific example satisfy the unambiguity requirement: The final word obtained uniquely determines the sequence of morphisms applied.

It is not necessary to begin with *two* morphisms h_0 and h_1. For instance, if one wants to use decimal notation in the plaintext, then it is natural to begin with ten morphisms and construct ten substitutions as well.

A special advantage of the public key cryptosystem discussed in Example 7.7 is that the setup "encryption key only" is not likely to lead into success in cryptanalysis: It is unlikely to find the secret decryption key from the public encryption key only. This state of affairs is due to the fact that there seems to be no way of recognizing a *candidate* for the decryption key to be the *correct* decryption key. (Indeed, any behavior of the actual letters can also be simulated by the junk letters.) Such a recognition is possible for the well-known cryptosystems discussed in the next two sections, so for these systems cryptanalysis based only on the encryption key might succeed with good luck!

On the other hand, an obvious disadvantage of the system discussed in Example 7.7 is that the cryptotext might become unreasonably long. The choice of the initial morphisms should be made to prevent this. On the other hand, the initial morphisms should be such that any sequence of them can be recovered (uniquely and fast) from the result of applying that sequence to the axiom w. Intuitively, these two requirements for the initial morphisms seem to be somewhat conflicting. It is an open problem, basically in the theory of DT0L systems, to exhibit a reasonably big class of systems satisfying these requirements.   □

Our final result in this section can be considered as an indication of how difficult it is to formalize the theory of cryptosystems, especially public key cryptosystems. We first list some reasonable requirements for public key cryptosystems. We then show that it is impossible to satisfy all these requirements simultaneously.

As already pointed out before, "easy" and "intractable" should be understood in the complexity-theoretic sense. (The details are not important, at least not for Theorem 7.1.) We are now ready to state the requirements.

A pair (E, D) is called a **skeletal public key cryptosystem** iff conditions (i)–(v) are satisfied.

(i) $E = \{e_i \mid i = 1, 2, \ldots\}$ and $D = \{d_i \mid i = 1, 2, \ldots\}$ are denumerable sets of algorithms. The algorithms e_i (resp. d_i) operate on inputs called *plaintexts* (resp. *cryptotexts*).

(ii) Given an arbitrary $i \geq 1$, the descriptions for the algorithms e_i and d_i are easily obtainable. For $i \neq j$, the descriptions for e_i and e_j, as well as those for d_i and d_j, are different.

(iii) Given the description for e_i and a plaintext w, the value $e_i(w)$ is easy to compute.

(iv) Given the description for d_i and a cryptotext c, the value $d_i(c)$ is easy to compute.

(v) For almost all i, it is intractable to obtain, from the description for e_i, the description for d_i or for any algorithm equivalent to d_i. (Equivalence means here that the input-output behavior is the same and that the complexity is essentially the same.)

Intuitively, the encryption and decryption keys of a public key cryptosystem should satisfy (i)–(v). We have used the term *skeletal* to indicate that conditions (i)–(v) are by no means intended to be exhaustive. We have not stated, for instance, the very essential requirement of e_i and d_i being inverses.

It is an immediate consequence of (ii) and (v) that, for almost all i, the computing of the index i from e_i is intractable. This is a rather strange requirement from the point of view of recursive function theory, for instance. (See Chapter 4.) Therefore, the following result is not surprising.

Theorem 7.1.   *There are no skeletal public key cryptosystems.*

Proof.   Assume the contrary: Some pair (E, D) satisfies (i)–(v). Given the description for an arbitrary e_i, we present the description for an algorithm d_i' equivalent to d_i. Clearly, this will contradict (v).

Our description for d_i' is as follows. Assume that c is an input for d_i'. Then d_i' first generates integers $j = 1, 2, \ldots$ and, by (ii), produces the description for e_j. Each description thus produced is compared with the given description for e_i until a description identical with the given description is found. (This will eventually happen.) Let the description found correspond to the index $j = k$. Finally, d_i' produces d_k by (ii) and computes the value $d_k(c)$ by (iv).

Clearly, for every input c, our algorithm d_i' produces the same output as d_i. Moreover, an upper bound for the complexity of d_i' is obtained by adding a constant to the complexity of d_i. This follows because the preprocessing done by d_i' is completely independent of the input c. Thus, from the point of view of complexity, preprocessing contributes only the same fixed amount for all inputs.   □

Theorem 7.1 should be understood in a proper perspective. It has nothing negative to say about the usefulness of public key cryptography or particular public key cryptosystems. One has to realize that the constant due to preprocessing might be $10^{10^{10}}$. The addition of such a constant makes an essential difference for all practical purposes.

On the other hand, Theorem 7.1 says that some ideas of computational complexity are hard to formalize in such a way that cryptographic security is taken into account. One can also say that asymptotic security cannot be obtained if the same key is used to encrypt arbitrarily long plaintexts. In fact, this concerns both classical and public key cryptosystems.

7.3. KNAPSACK SYSTEMS

In the next two sections we discuss the two most common and widely studied types of public key cryptosystems: knapsack and RSA systems.

The language KNAPSACK was introduced in Example 6.2. In cryptography it is convenient to keep the "size of the knapsack" as a separately given quantity. Thus, we define an **n-dimensional knapsack vector** to be an n-tuple $(a_1, a_2, \ldots, a_n)$ of distinct positive integers.

A knapsack vector $(a_1, a_2, \ldots, a_n)$ is used as an **encryption key** as follows. The plaintext is expressed in bits and divided into blocks of n bits each. Each block B is viewed as an n-dimensional column vector (consisting of 0s and 1s) and is encrypted as the number

$$c = (a_1, a_2, \ldots, a_n)B. \tag{6}$$

If the encryption key and the cryptotext are known, cryptanalysis still amounts to solving the knapsack problem that is known to be an $\mathfrak{NP}$-complete problem.

Observe that here the knapsack problem is understood in the form: Which of the numbers a_i sum up (exactly) to c? In Chapter 6 the problem was understood in the form: Do some of the numbers a_i actually sum (exactly) to c? Both versions are $\mathfrak{NP}$-complete. Observe also that the sorting of numbers can be done quickly, so it does not make much difference from the complexity point of view whether or not the vectors a_i in a knapsack vector are sorted in an increasing order. Finally, observe that the encryption of (6) is not necessarily one-to-one; several plaintexts may correspond to the same cryptotext c because the knapsack problem determined by c and the vector $(a_1, \ldots, a_n)$ may have several solutions.

In a **knapsack system**, a knapsack vector constitutes the **public encryption key**. The key is used for encryption as described above. We now describe the trapdoor information for the legal recipient of messages.

A knapsack vector $(a_1, a_2, \ldots, a_n)$ is called **super-increasing** iff each number exceeds the sum of the preceding numbers—that is,

$$a_j > \sum_{i=1}^{j-1} a_i \qquad \text{for } j = 2, \ldots, n. \tag{7}$$

In spite of the difficulty of the general knapsack problem, the special case dealing with super-increasing knapsack vectors is easy.

Lemma 7.2. Let $\mathbf{A} = (a_1, \ldots, a_n)$ *be a super-increasing knapsack vector and c a positive integer. Then the resulting knapsack problem (that is, the problem of the existence of numbers a_i whose sum equals c) is solvable in linear time. Moreover, if a solution exists, it is unique.*

Proof. Denote by $(m_1, \ldots, m_n)^T$ a possible solution (column) vector. It is an immediate consequence of (7) that

$$c \geq a_n \quad \text{iff} \quad m_n = 1. \tag{8}$$

This condition can be extended to concern the other components as well:

$$c - \sum_{j=i+1}^{n} m_j a_j \geq a_i \quad \text{iff} \quad m_i = 1, \tag{9}$$

for $i = n - 1, \ldots, 1$. By (8), the value of m_n is uniquely determined by c. By an inductive argument based on (9), we conclude that there is at most one solution vector $(m_1, \ldots, m_n)^T$. Moreover, the complexity result contained in Lemma 7.2 follows because the eventual solution vector is found by one "sweep" from right to left, using (8) and (9). $\square$

Consider now a super-increasing knapsack vector $(a_1', a_2', \ldots, a_n') = \mathbf{A}'$. Choose an integer $M > 2a_n'$, referred to as the modulus. Another integer u, with no common factors with M, is chosen. Consequently, the inverse u^{-1} with the property $uu^{-1} \equiv 1 \pmod{M}$ exists (and, in fact, can be found very fast).

Next, "scramble" the given vector $\mathbf{A}'$ by multiplying the components by u and reducing the products modulo M. In other words, form a new knapsack vector $(a_1, a_2, \ldots, a_n) = \mathbf{A}$ where a_i is the least positive residue of ua_i' $\pmod{M}$.

Lemma 7.3. *Assume that $\mathbf{A}'$, M, u, and $\mathbf{A}$ are defined as above. Let c' be a positive integer and let c be the least positive residue of uc' modulo M. Then the knapsack problem determined by $\mathbf{A}'$ and c' has at most one solution. Similarly, the knapsack problem determined by $\mathbf{A}$ and c has at most one solution. If a solution to the latter problem exists, it equals the unique solution of the former problem.*

Proof. It is an immediate consequence of Lemma 7.2 that the solution X' of the problem determined by $\mathbf{A}'$ and c' is unique.

Assume that X is some solution of the problem determined by $\mathbf{A}$ and c. Hence, $c = \mathbf{A}X$. From this, we obtain

$$c' \equiv u^{-1}c = u^{-1}\mathbf{A}X \equiv u^{-1}u\mathbf{A}'X \equiv \mathbf{A}'X \pmod{M}.$$

Since $\mathbf{A}'$ is super-increasing, the choice of M implies that $\mathbf{A}'X < M$ and, consequently, $c' = \mathbf{A}'X$.

Thus, we have shown that any solution of the problem determined by $\mathbf{A}$ and c is a solution of the problem determined by $\mathbf{A}'$ and c'. Since the latter problem possesses at most one solution, so does the former problem. We have now also established the last sentence of the lemma. $\square$

It is possible that the knapsack problem determined by A' and c' has a solution X', but the problem determined by A and c has no solutions. (This may happen when X' is only a "modular" solution of the latter problem.) However, such a possibility is irrelevant in the following discussion.

We are now ready to fill in the missing details concerning knapsack cryptosystems. The public encryption key is a knapsack vector A, obtained from a super-increasing vector A' by a modular multiplication in the way described above. The secret **trapdoor information** consists of the items A', u, u^{-1}, and M.

When the legal recipient possessing the trapdoor information receives the cryptotext c, he or she first computes the least positive remainder c' of $u^{-1}c$ modulo M and then solves the knapsack problem determined by A' and c'. By Lemma 7.2, this can be done quickly. Lemma 7.3 guarantees that the plaintext obtained is the original one. Observe that the last sentence of Lemma 7.3 can be applied, since we *know* that the problem determined by A and c has a solution—c was constructed directly from a solution.

Example 7.8. Consider the following ten-dimensional super-increasing knapsack vector:

$$A' = (1, 3, 5, 11, 21, 44, 87, 175, 349, 701).$$

Choose the (sufficiently large) modulus $M = 1590$ and multiplier $u = 43$. Because $43 \cdot 37 = 1591 \equiv 1 \pmod{1590}$, we conclude that $u^{-1} = 37$.

Modular multiplication by u now yields the following knapsack vector, used as the public encryption key:

$$A = (43, 129, 215, 473, 903, 302, 561, 1165, 697, 1523).$$

Hence, the plaintext has to be divided into blocks consisting of ten bits.

Clearly, five bits are needed to represent one letter of the English alphabet. This means that blocks of two letters will be encrypted simultaneously.

To represent the ith letter, we write i in the binary notation. Again, we consider A to be the 0th letter (not to be confused with the vector A!), B the first letter, and so forth. Thus, 00000 represents A, 00010 represents C, and 10001 represents R.

The bit representation of the plaintext CRYPTO is:

$$0001010001, 1100001111, 1001101110. \tag{10}$$

We have divided the entire sequence of 30 bits into blocks consisting of 10 bits each.

The blocks are encrypted separately, using the vector A. Thus, the encryption of the first block is

$$473 + 302 + 1523 = 2298.$$

Similarly, the two other blocks are encrypted as 4118 and 3842. Thus, the whole cryptotext is the triple

$$(2298, 4118, 3842).$$

To decrypt, we first multiply the items of the triple by $u^{-1} = 37$ and reduce the products modulo 1590, obtaining the triple

$$(756, 1316, 644). \tag{11}$$

We then solve the knapsack problem determined by $\mathbf{A}'$ and 756 by the method of Lemma 7.2. The last component of the solution will be 1 because $756 \geq 701$. The next three components, counting from the end, will be 0 because, otherwise, 756 is exceeded. However, the number 44 is all right, and so is the number 11. Hence, the whole solution is the column vector $(0001010001)^T$.

When the other two items of the triple (11) are treated similarly, we recover the plaintext of (10).

Thus, the legal recipient is able to decrypt quickly. This example is much too simple for cryptanalysis to be very difficult. In fact, the first few entries of $\mathbf{A}$ suggest (correctly!) the multiplier used. □

Many variants of knapsack cryptosystems have been suggested. Indeed, the knapsack problem seems to constitute an ideal basis for a public key cryptosystem. The resulting encryption and decryption (based on trapdoor information) techniques are really simple—for instance, considerably simpler than in connection with the RSA system discussed in the next section. An $\mathfrak{NP}$-complete problem, or a problem "looking like" an $\mathfrak{NP}$-complete problem, has to be attacked in connection with cryptanalysis. On the negative side is the famous result due to Adi Shamir (and discussed below) that the cryptanalysis of an essential subcase of knapsack systems can be carried out in polynomial time.

The suggested variants of knapsack cryptosystems seem to meet the two main requirements of cryptography, privacy and signature generation, fairly well. Since these two requirements are in some sense conflicting, it seems hard to satisfy both in a really strong fashion by the same system. The basic variant of a trapdoor knapsack system just discussed satisfies the requirement of privacy. We shall now discuss a variant especially suitable for generating signatures. The emphasis is on speed and simplicity: Both the signing and the verification can be done by performing only additions and subtractions.

We need the following variant of the knapsack problem, easily shown to be $\mathfrak{NP}$-complete.

Given $n + 2$ integers $a_1, \ldots, a_n, k$, and m, find (if possible) some solution $(c_1, \ldots, c_n)$ for the congruence

$$m \equiv \sum_{j=1}^{n} c_j a_j \pmod{k}, \tag{12}$$

where each c_j is a small integer (more specifically, $0 \le c_j \le \log_2 k$).

Before proceeding with the formal details, we discuss in general terms how such a knapsack system can be used to generate signatures. The other party A chooses and publicizes a knapsack system determined by $a_1, \ldots, a_n$, k such that the system is apparently difficult but can actually be solved quickly by some secret trapdoor information. Party A signs a message m by using the trapdoor information for solving (12): the n-tuple $(c_1, \ldots, c_n)$ will constitute his or her signature for m. The recipient B (who received both m and the signature $(c_1, \ldots, c_n)$) can verify the signature by checking that (12) holds. If B or some other party C wants to forge A's signature on a message m', he or she has to solve the corresponding instance of the knapsack problem. (A special requirement concerning the chosen knapsack system is that all conceivable messages m must have a signature—that is, (12) must have a solution for all such m.)

We are now ready to present the formal details. Consider a prime number k whose binary representation possesses t bits. (Typically, $t = 100$.) Let $H = (h_{ij})$ be a $t \times 2t$ matrix whose entries are randomly chosen bits. Let $\mathbf{A}$ be a $2t$-dimensional column vector satisfying the following t congruences:

$$\mathbf{HA} \equiv \begin{pmatrix} 2^0 \\ 2^1 \\ \vdots \\ 2^{t-1} \end{pmatrix} \pmod{k}. \tag{13}$$

Observe that there are only t congruences in $2t$ unknowns in (13). We may, thus, basically choose t components of $\mathbf{A}$ at random and compute the remaining components. The computation can be done fast and the probability of getting stuck is small, especially because we can alter some of the randomly chosen components whenever necessary.

The components a_j of $\mathbf{A}$ will be random-looking t-bit integers such that any power of 2 between 2^0 and 2^{t-1} can be expressed (mod k) as the sum of some of them.

The items $\mathbf{A}$ and k are now publicized, whereas H is kept as a secret trapdoor information. Messages m are numbers in the closed interval [0, $k - 1$]. The signature for m is a vector $(c_1, \ldots, c_{2t}) = \mathbf{C}$ satisfying (12). (Observe that now $n = 2t$.)

Signatures can immediately be verified by checking (12). Forging of signatures for messages m' will be difficult because of reasons explained above. (Essentially, you have to solve the $\mathcal{NP}$-complete modular knapsack problem.)

On the other hand, signing will be easy if we are in the possession of the secret trapdoor information H. In order to sign a message m, we write m as a sum of powers of 2:

$$m = \sum_{i=0}^{t-1} b_i 2^i,$$

where b_i is the ith bit, counted from the right, in the binary representation of m. Since the powers of 2 involved can be expressed, by (13), in terms of the components of $\mathbf{A}$, we obtain

$$m = \sum_{i=0}^{t-1} b_i 2^i \equiv \sum_{i=0}^{t-1} b_i \sum_{j=1}^{2t} h_{ij} a_j$$

$$= \sum_{j=1}^{2t} \left(\sum_{i=0}^{t-1} b_i h_{ij} \right) a_j = \sum_{j=1}^{2t} c_j a_j \pmod{k}.$$

Clearly, the numbers c_j are obtainable (fast!) from the trapdoor information. Moreover, it is immediately seen that they lie in the range required in (12).

The most immediate insecurity in this signing procedure is due to an argument similar to the one presented in connection with Hill's system, Example 7.3. Whenever sufficiently many message-signature pairs are known, the matrix H can be computed from them by the methods of linear algebra.

This insecurity problem can be solved by randomizing the bits of m before signing m. This can be done, for instance, by subtracting a randomly chosen subset of the a_j's from m:

$$m' \equiv m - \sum_{j=1}^{2t} r_j a_j \pmod{k},$$

where $(r_1, \ldots, r_{2t}) = \mathbf{R}$ is a random vector of bits. We first find a signature $\mathbf{C}'$ for m' by the method described above. Then $\mathbf{C}' + \mathbf{R}$ can be used as the signature for the original m because

$$m \equiv m' + \mathbf{R}\mathbf{A} \equiv \mathbf{C}'\mathbf{A} + \mathbf{R}\mathbf{A} = (\mathbf{C}' + \mathbf{R})\mathbf{A} \pmod{k}.$$

In the very unlikely event that some of the components of $\mathbf{C}' + \mathbf{R}$ exceed $\log_2 k$ (recall the range of the number c_j in (12)), we have to try again with another random vector $\mathbf{R}$. (In fact, a more-detailed analysis will show that a new choice of $\mathbf{R}$ is never necessary.)

We now discuss possibilities of "breaking" the knapsack system —that is, carrying out successful cryptanalysis in polynomial time.

Although a cryptosystem is based on an $\mathfrak{NP}$-complete problem, possibilities for fast cryptanalysis still exist. There might be various ways of avoiding the $\mathfrak{NP}$-complete problem. For instance, several plaintext-crypto-text pairs may reduce cryptanalysis to an easy problem in linear algebra. Examples of this were given earlier. Moreover, in regard to trapdoor knapsack systems, the following warning should also be kept in mind. The special case of the knapsack problem, where the knapsack vector results from a super-increasing one by some modular multiplication, has not been shown to be $\mathfrak{NP}$-complete. This special case might look like the general knapsack problem—especially if some further precaution such as permuting the components of the resulting vector **A** and keeping the permutation secret, is taken. However, it is only a special case of the general problem. One reason for this is that, by Lemma 7.3, solutions are always unique in the special case—a property not shared by the general knapsack problem.

In spite of these facts, the cryptanalysis of trapdoor knapsack systems remained an open problem for quite a long time, until A. Shamir succeeded in breaking a basic variant of the systems in 1982. His algorithm is outlined in Example 7.9.

This result should by no means be considered as the death of knapsack systems. As pointed out earlier, the knapsack problem is in many ways a very suitable basis for public key cryptosystems. Cryptanalytic attacks can always be defended through some new safety measures. Because of the very nature of cryptography, it is rather difficult to obtain any "final" results in the field. As emphasized by Shamir, cryptography remains a never-ending struggle between code breakers and code makers—that is, between cryptanalysts and designers of cryptosystems.

Example 7.9. We face the following cryptanalytic task. We know a knapsack vector $\mathbf{A} = (a_1, \ldots, a_n)$, which is used as a public encryption key in the manner described above. We also know that **A** has been obtained from some super-increasing knapsack vector $\mathbf{A}'$ by modular multiplication, using some multiplier u and modulus M, but $\mathbf{A}'$, u, and M are unknown to us. We want to find u^{-1} and M, after which we can recover $\mathbf{A}'$ by multiplying **A** with u^{-1} and reducing the products modulo M.

Observe that the cryptanalytic setup here is *encryption key only*. In general, this means that more time is available because the analysis of the system can be carried out before important cryptotexts have been sent.

The algorithm outlined next works in polynomial time with respect to the size of the given knapsack vector **A**. We thus have to consider a family of knapsack vectors whose sizes grow to infinity. There are two parameters contributing to the size of a knapsack vector **A**: the number n and the sizes of the components a_i. If either one of the parameters is kept bounded from above, the resulting knapsack problems can be solved trivially in polynomial time. Thus, both must be unbounded. We assume that in the class of knapsack vectors **A** considered, the sizes of the components a_i—and, hence, the secret modulus M—grow linearly with n.

The initial observation is that it is not necessary for us to find the multiplier u^{-1} and modulus M actually used by the designer of the cryptosystem. Any multiplier w and modulus M will do as well, provided the knapsack vector $\mathbf{A}'$ resulting from $\mathbf{A}$ is super-increasing and M is sufficiently large with respect to $\mathbf{A}'$. Such pairs (w, M) are referred to as **trapdoor pairs.** If we have found a trapdoor pair, Lemma 7.3 becomes available, and we may decrypt using the resulting super-increasing knapsack vector. This is quite independent of whether or not our trapdoor pair and the resulting super-increasing vector $\mathbf{A}'$ are the ones actually used by the cryptosystem designer. On the other hand, the existence of at least one trapdoor pair is guaranteed by the fact that the cryptosystem designer made use of such a pair.

To find a trapdoor pair (w, M), we first consider the graphs of the functions $a_i w \pmod{M}$ for all values $i = 1, \ldots, n$. Here w is understood as a variable ranging over reals rather than over integers. Clearly, the graph of $a_i w \pmod{M}$ has the sawtooth form of Figure 7.1. This sawtooth curve is considered for each value of $i = 1, \ldots, n$.

Since $a_i w \pmod{M}$ has to be small in comparison with M (recall that the numbers $a_i w$ form a super-increasing knapsack vector and M is large in comparison with this vector), we conclude that the trapdoor pair value of w must be close to some a_i-minimum. This holds true for all values of i. Consequently (and this is a very important conclusion), there must be some small interval containing an a_i-minimum, for all i. From this interval we also find the value of w.

We now come to the problem of how to express these ideas in terms of inequalities. The first obstacle is that we do not know the value of the modulus M. This obstacle is easily overcome: We just reduce the size of the above picture so that M is replaced by 1. In other words, the lengths are

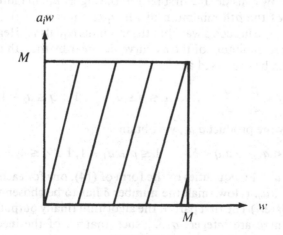

FIGURE 7.1

divided by M. It is obvious that this operation does not affect the location of the *accumulation point* for the a_i-minima in which we are interested. For instance, if there was an a_i-minimum near by the seventh a_3-minimum for all i before the size reduction, the same certainty holds true after the size reduction.

The algorithm for finding a trapdoor pair consists of two parts. In the first part, we find candidates for an integer p such that the pth minimum of the sawtooth curve corresponding to a_1 is the accumulation point we are looking for. The second part of the algorithm tests the candidates one after the other. Because we know that an accumulation point exists, the algorithm terminates.

A specific precaution has to be taken. The first part of the algorithm might produce too many (in comparison with the size of the problem) candidates for p. Therefore, we fix—in advance—the value for the parameter k indicating the maximum number of candidates allowed. If the first part of the algorithm produces $k + 1$ candidates for p, the algorithm terminates and reports failure.

On the other hand, in the first part of the algorithm we do not have to consider all the possible components $a_2, \ldots, a_n$, but we may fix in advance the value for another parameter $u < n$ and consider only the components $a_2, \ldots, a_u$. In other words, the first part of the algorithm produces numbers p such that the pth minimum of a_1 is nearby some minimum of a_i, for every $i = 2, \ldots, u$. Thus, the values of $i > u$ are not considered at all in the first part of the algorithm. Hence, the first part might produce entirely wrong values of p. On the other hand, the second part of the algorithm checks all values of i in the interval $2 \leq i \leq n$: A candidate p is immediately discarded if, for some i, there is no minimum of the sawtooth curve corresponding to a_i near the pth minimum of the curve corresponding to a_1.

Let us now consider the first part of the algorithm in more detail. The w-coordinate of the pth minimum of the a_1-curve is p/a_1. (Recall that we reduced the picture in such a way that the modulus equals 1.) Hence, the condition that some minimum of the a_2-curve lies near by the pth minimum of the a_1-curve can be expressed as

$$-\epsilon < \frac{p}{a_1} - \frac{q}{a_2} < \epsilon, \qquad 1 \leq p \leq a_1 - 1, 1 \leq q \leq a_2 - 1.$$

Multiplying by the product $a_1 a_2$, we obtain

$$-\delta < a_2 p - a_1 q < \delta, \qquad 1 \leq p \leq a_1 - 1, 1 \leq q \leq a_2 - 1. \tag{14}$$

We now write $u - 1$ inequalities of the form of (14), one for each of the components $a_2, \ldots, a_u$. (How small the number δ has to be chosen will be commented upon later.) The first part of the algorithm finally outputs all integers p for which there are integers $q, \ldots$ such that all of the inequalities are satisfied.

We now describe the second part of the algorithm. It tests the numbers p produced by the first part until it is successful. At first all discontinuity points (of all n curves) lying in the interval

$$\left[\frac{p}{a_1}, \frac{p+1}{a_1} \right]$$

are found and sorted into increasing order. Let x_t and x_{t+1} be two consecutive points in the sorted list of points. Then, in the interval $[x_t, x_{t+1}]$ each of the a_i-curves is just a line segment, expressible in the form

$$a_i w - c_i^t,$$

where c_i^t is a constant depending on i and t (and, of course, also on p).

The solution of the following system of linear inequalities in w is a (possibly empty) subinterval of $[x_t, x_{t+1}]$:

$$x_t \leq w \leq x_{t+1}, \tag{15}$$

$$\sum_{i=1}^{n} (a_i w - c_i^t) < 1, \tag{16}$$

$$a_i w - c_i^t > \sum_{j=1}^{i-1} (a_j w - c_j^t), \qquad i = 2, \ldots, n. \tag{17}$$

Moreover, a necessary and sufficient condition for two numbers w and M to constitute a trapdoor pair is the membership of w/M in some such subinterval. Indeed, (17) expresses the super-increasing condition and (16) expresses the condition for the modulus to be sufficiently large.

Thus, the second part of the algorithm investigates successively through pairs (p, t), where p is a candidate produced by the first part and t is an index of a point in the sorted list corresponding to p. The investigation is carried out until a nonempty interval is found. The termination is guaranteed by the fact that the trapdoor pair actually used by the cryptosystem designer corresponds to a nonempty interval. This completes our description of the algorithm.

Since linear integer programming lies beyond the scope of this book, we are not in a position to deduce detailed complexity estimates for the algorithm. Basically, inequalities (14)–(17) can be handled very fast and certainly in polynomial time, even in the worst case. Moreover, fundamental estimates concerning the probability of the success of the algorithm can also be deduced. "Success" here means that the algorithm does not produce (in its first part) too many candidates for p. Recall that, before running the algorithm, we fixed the values of two parameters k and u. Here k is the highest-permissible number of candidates for p, and u is the number of components studied in the first part of the algorithm. Assuming that δ in (14) is chosen to be less than $\sqrt{a_1}/2$ and u is at least 3, then the probability for the algorithm failing is at most

$$\left(\frac{2}{k} \right)^{u-1}. \tag{18}$$

For instance, when allowing $k = 100$ candidates, even the value $u = 3$ (that is, considering only two inequalities in the first part of the algorithm) induces a rather small probability of failure. Of course, both the increasing of the allowable number of candidates for p and the increasing of the number of the a_i-components studied in the first part of the algorithm are likely to decrease the estimated probability for failure. This is seen also from the estimate in (18). $\qquad\qquad\qquad\qquad\qquad\qquad\qquad\qquad\qquad\qquad\qquad\qquad$ □

As we have already emphasized, knapsack systems possess many desirable features from the point of view of cryptography. Example 7.9 is to be understood as one illustration of the struggle between cryptanalysts and cryptosystem designers. There are various ways of making a knapsack cryptosystem more secure. For instance, it is easy to see that trapdoor knapsack systems are not closed under composition: If you begin with a super-increasing vector and perform two successive modular multiplications, then the resulting vector is not in general obtainable from the original vector by only one modular multiplication. Our final example in this section discusses compositions of knapsacks in this sense.

Example 7.9a.   The idea of having a super-increasing knapsack vector as a starting point when designing cryptosystems is in accordance with the trapdoor techniques discussed in Section 7.2. The problem associated with a super-increasing vector is an easy subproblem of the general intractable knapsack problem.

However, we show now that in knapsack-based cryptosystem design, it is not necessary to begin with an easy knapsack. In fact, we may begin with an arbitrary knapsack.

Let $(a_1^1, \ldots, a_n^1)$ be an arbitrary knapsack vector. We perform $n - 1$ modular multiplications, obtaining a new vector each time. Assume we have already obtained the vector $(a_1^j, \ldots, a_n^j)$, where $1 \le j \le n - 1$. We choose a multiplier u^j and a modulus M^j such that they have no common factors and the modulus is large enough:

$$M^j > \sum_{i=1}^{n} a_i^j. \tag{19}$$

The number a_i^{j+1}, $i = 1, \ldots, n$, is now defined to be the smallest positive remainder of $u^j a_i^j \pmod{M^j}$.

The knapsack vector $(a_1^n, \ldots, a_n^n)$ constitutes the public encryption key, to be used as in connection with other knapsack systems. Everything else is kept secret, in particular the multipliers and moduli.

The legal recipient (who knows the secret information) can now decrypt a cryptotext b^n by standard linear algebra as follows. To find the bits x_i from the equation

$$\sum_{i=1}^{n} x_i a_i^n = b^n,$$

the equation is first multiplied by the inverse of u^{n-1}, yielding for some b^{n-1} the congruence

$$\sum_{i=1}^{n} x_i a_i^{n-1} \equiv b^{n-1} \pmod{M^{n-1}}.$$

However, (19) implies that this congruence is, in fact, an equation. Multiplying the equation by the inverse of u^{n-2} and continuing in the same way, the following system of linear equations is obtained:

$$\sum_{i=1}^{n} x_i a_i^{j} = b^{j}, \qquad j = 1, \ldots, n.$$

The probability of the system being nonsingular is high, and the x_i's can be quickly found.

7.4. RSA SYSTEM

The other widely studied public key cryptosystem was originally introduced by Rivest, Shamir, and Adleman and is usually referred to as RSA. It is based on the fact that it is almost impossible to recover two large primes p and q from their product $n = pq$—at least according to the presently known factorization algorithms. On the other hand, large random primes can be generated quickly.

The public key encryption is based on n, whereas decryption requires that we know p and q. We now present the details.

Let p and q be two large random primes (typically, having at least 100 digits in their decimal representation). Let

$$n = pq \quad \text{and} \quad \varphi(n) = (p-1)(q-1).$$

(In fact, φ is the well-known Euler function.) Choose a number $e > 1$ relatively prime to $\varphi(n)$ and a number d satisfying the congruence

$$ed \equiv 1 \pmod{\varphi(n)}. \tag{20}$$

(Since e and $\varphi(n)$ are relatively prime, the congruence has a solution. It can be found rapidly by Euclid's algorithm.)

The numbers n and e, referred to as the **modulus** and the **encryption exponent**, respectively, constitute the public encryption key. Basically, the cryptotext c will be the least positive remainder of $w^e \pmod{n}$, where w is the plaintext.

More specifically, we first express the plaintext as a word over the alphabet $\{0, 1, 2, \ldots, 9\}$. The word is divided into blocks of suitable size. The blocks are encrypted separately by applying the pair (n, e) in the way described earlier. A suitable size of the blocks is the unique integer i satisfying the inequalities $10^{i-1} < n < 10^{i}$.

Cryptanalysis of the RSA system will be discussed later on. We now show that decryption is easy if you are in the possession of the secret trapdoor information—that is, the number d, referred to as the **decryption exponent**. At this point, the knowledge of d is already intimately connected with the knowledge of p and q. The next result shows that d is, indeed, a proper decryption exponent.

Theorem 7.4. *Let c be the cryptotext corresponding to a plaintext w obtained in the RSA way described above, and let d satisfy (20). Then $c^d \equiv w$ (mod n).*

Proof. By (20), there exists an integer j such that $ed = j\varphi(n) + 1$. According to Euler's well-known theorem,

$$w^{\varphi(n)} \equiv 1 \;(\mathrm{mod}\; n),$$

assuming that neither p nor q divides w. Hence, under this same assumption,

$$w^{j\varphi(n) + 1} \equiv w \;(\mathrm{mod}\; n). \tag{21}$$

On the other hand, it is easy to see that (21) also holds true in the case where at least one of p and q divides w. Hence, (21) is universally valid.

Consequently, we obtain

$$c^d \equiv (w^e)^d \equiv w^{j\varphi(n) + 1} \equiv w \;(\mathrm{mod}\; n). \qquad \square$$

Example 7.10. We choose $p = 47$ and $q = 59$, yielding $n = 2773$ and $\varphi(n) = 2668$. We further choose $e = 17$, yielding $d = 157$. Our earlier remark applies here also: The numbers chosen are much too small. However, we want the reader to be able to follow the computations. Thus, (n, e) constitutes the public encryption key.

We want to carry out the encryption in detail, starting from a plaintext in a natural language. In fact, another dimension of security can be provided in all cryptosystems by choosing the plaintext language to be one not too widely understood. Our plaintext language here is Finnish, and we consider the plaintext TURUSSA TUULEE (meaning "Turku is a windy place"). The alphabet will be the one used in Example 7.2. The letters will be numbered using two digits from the "decimal alphabet" $\{0, 1, \ldots, 9\}$. However, we now reserve the number 00 for the empty space between two plaintext words. Thus, the letter A gets the number 01.

Since $10^3 < n < 10^4$, we conclude that the plaintext blocks will consist of 4 digits. Thus, the numerical encoding of the plaintext will be:

$$1920 \quad 1720 \quad 1818 \quad 0100 \quad 1920 \quad 2012 \quad 0505.$$

From this we obtain the cryptotext

$$2109 \quad 0538 \quad 1992 \quad 1952 \quad 2109 \quad 1453 \quad 2390.$$

We may still check that decryption works:

$$2109^{157} \equiv 1920, \; 2390^{157} \equiv 505 \pmod{2773},$$

and so on.

The reader is surely able to factor 2773 and, hence, to find the decryption exponent $d = 157$. The reader might still want to consider the case where p and q are so large that they cannot be recovered from their product n. How could cryptanalysis be carried out in this case?   □

Indeed, no cryptanalytic attack against the RSA cryptosystem has so far been successful. Factoring a number seems to be much harder than determining whether the number is prime or composite.

Even if one admits that factoring is intractable, it would still be conceivable that $\varphi(n)$ (and, hence, d) could be determined by some "direct" approach. However, if this were the case, we would then immediately obtain the following fast-factoring algorithm.

Assume, thus, that we have been able to compute $\varphi(n)$. Then $p + q$ is immediately obtainable from $\varphi(n) = n - (p + q) + 1$ and the public information n. On the other hand, $p - q$ is the square root of $(p + q)^2 - 4n$ and, thus, also immediately obtainable. Finally, q can be immediately computed from $p + q$ and $p - q$.

Thus, an algorithm for computing $\varphi(n)$ immediately yields an algorithm of essentially the same complexity for factoring n. An example of a case where the cryptographic security of the RSA system is provably equivalent, from the point of view of complexity, to the security of keeping the factors of n secret is given in Example 7.11. This gives an indication of the security of the RSA system.

A much-stronger result, pointing towards the cryptographic security of the RSA system, has been obtained recently, [Be-OCS]. In general, it is a major problem concerning all cryptosystems that some useful information might be obtainable from the cryptotext, although the complete breaking of the text has not been successful. One still might be able to recover some crucial bit of the plaintext, or at least some partial information of the plaintext. A concrete example concerning RSA systems is the case where the last digit of n is 3. Then it immediately follows that the last digits of p and q are either 1 and 3 or else 7 and 9. It is conceivable that in some cases this information could be very useful. The significance of such partial information to the cryptographic security of the cryptosystem is far from being understood.

The results established in [Be-OCS] tell, essentially, that even a partial success in the cryptanalysis of the RSA system would lead to a complete breaking of the system. Such partial success could be the computation of a certain bit of the plaintext or the guessing of the last bit of the plaintext with an error probability less than $\frac{1}{4} - \epsilon$.

When RSA systems are used for signatures, there are a number of open problems similar to ones appearing in connection with all public key cryptosystems. For instance, some messages could be signed without the trapdoor information. The number of such messages might become very large if sufficiently many previously signed messages are known. Results like these depend on the arithmetical properties of n.

The following example gives an explicit interconnection between factorizing n and cryptanalysis.

Example 7.11. We now consider the system usually referred to as **Rabin's system,** where the encryption exponent e equals 2. In this case the remarkable result can be obtained that cryptanalysis is actually equivalent to factoring n. Here the setup for cryptanalysis is *cryptotext only,* and equivalence is understood in the complexity-theoretic sense: An algorithm for solving one of the problems yields an algorithm of essentially the same complexity for solving the other problem. We now outline the argument showing the complexity-theoretic equivalence.

Consider the RSA system with the public key $(n, 2)$. If the cryptanalyst knows the factorization of n, then he or she is able to compute $\varphi(n)$ and obtain, by Euclid's algorithm, the decryption exponent d.

There is also a fast algorithm for solving the congruence

$$x^2 \equiv c \,(\mathrm{mod}\, n), \tag{22}$$

provided the factorization of n is known. Of course, this gives a method of finding the plaintext from the cryptotext.

Assume, conversely, that the cryptanalyst is able to decrypt. This assumption can be formalized by saying that the cryptanalyst is in the possession of an oracle M such that, given an input (c, n), M produces a solution x of the congruence (22). Here it is assumed that n is the product of two primes and that the congruence has solutions; that is, c is a quadratic residue (mod n). The latter assumption is immediately satisfied in the case considered because we know c is the result of encryption with the exponent 2.

If (22) is solvable, it has four incongruent solutions α, β, $n - \alpha$, and $n - \beta$. It is assumed that M randomly produces one of the solutions.

On the other hand, if we know two solutions α and β (such that $\beta \neq n - \alpha$), then we can factor n as follows. Since both α and β are solutions of (22), we obtain

$$\alpha^2 - \beta^2 \equiv 0 \,(\mathrm{mod}\, n).$$

This congruence tells us that the product pq divides the product $(\alpha + \beta)(\alpha - \beta)$. On the other hand, it follows from our assumptions ($\alpha \neq \beta$ and $\beta \neq n - \alpha$) that pq divides neither $\alpha + \beta$ nor $\alpha - \beta$. Hence, p must divide one of the factors of $\alpha^2 - \beta^2$, whereas q divides the other. But this means that the greatest common divisor of n and $\alpha - \beta$ is either p or q. Because the greatest common divisor can be computed quickly by Euclid's algorithm, we are able to factor n if we know α and β.

We now proceed as follows. We randomly choose a number α between 1 and n and give the oracle M the input consisting of $c = \alpha^2$ and n. Then M produces—with probability ½—one of the other solutions β and $n - \beta$. If this happens, we are able to factor n. Otherwise, we choose a new α and repeat the procedure. The probability of success converges to 1.

Algorithms like the one just described are commonly used in cryptography: They are fast, at least polynomial time, but there is a small probability of failure. The terms *Monte Carlo* and *Las Vegas* are used to indicate different types of failure. A **Monte Carlo algorithm** might give a wrong answer in some cases. A **Las Vegas algorithm** always gives a correct answer, but it might end up with the answer "I don't know" in some cases.

Clearly, the algorithm described is of the Las Vegas type. If M has not been able to produce the other solution of the congruence in any of the instances considered, then an "I don't know" answer results for the question of factoring n.   □

We conclude this section with a summary about known results concerning the complexity of primality testing and another number-theoretic problems important in cryptography. As we already pointed out, primality testing is essentially easier than factoring, at least according to our present knowledge.

The algorithms of Rabin and Solovay-Strassen are polynomial time but probabilistic of the Monte Carlo type. The algorithm of Miller works in deterministic polynomial time. It is, however, based on the assumption that the extended Riemann hypothesis holds true. Finally, Adleman's algorithm is also deterministic and is not based on any assumptions. However, it is still unsettled whether the algorithm is in polynomial time.

Hence, it is still open whether or not primality testing is in $\mathcal{P}$. For numbers of reasonable size (say, up to 1000 decimal digits), primality testing can be carried out by current algorithms and computers. At the time of this writing, the largest known prime number is $2^{86,243} - 1$.

Another number-theoretic problem with numerous applications concerns the solutions of the congruence

$$x^e \equiv c \pmod{n}, \tag{23}$$

where e, c, and n are given integers and, moreover, n is square-free. In most cases, n is either a prime or a product of two distinct primes. Various modifications of this problem are important from the point of view of cryptography.

A fast algorithm was known already to Gauss for settling the problem of whether or not (23) possesses a solution, given e, c, and a prime modulus,

n. No polynomial-time algorithm is known for this problem in case of a composite modulus n. (The factorization of n is not known.)

A somewhat different problem is to assume that (23) possesses a solution and study algorithms for finding a solution. If n is prime (or the factorization of n is known), then a random polynomial-time algorithm can be constructed. Example 7.11 shows that, for $e = 2$ and n as in RSA systems, the problem of solving (23) is equivalent to the problem of factorizing n. If this result could be extended to concern the general form of (23), this would show the security of the RSA system, under the assumption that the factorization of n is intractable.

7.5. PROTOCOLS FOR SOLVING SEEMINGLY IMPOSSIBLE PROBLEMS IN COMMUNICATION

Public key cryptography has considerably changed our ideas about what is impossible when several parties, adversaries or not, are communicating with each other. For instance, assume that two people want to find out who is older, without giving away any other information about their ages. Could a conversation between them take place in such a way that this requirement will be satisfied? Or could a person flip a coin to another person by telephone without any assisting third party? Intuitively, such tasks seem impossible. However, the use of one-way functions of public key cryptography makes the design of protocols for solving such problems possible.

We do not give a formal definition for the notion of a *protocol*. Essentially, a **protocol** is an algorithm to be followed by the different communicating parties. Protocols based on cryptography make substantial use of some one-way functions.

First consider the following very general task. A private conversation should be established between two individual users of an information system or a communication network. No assumptions are made concerning whether or not the two individual users ever communicated before.

Clearly, the basic idea behind public key cryptography can be used to solve this problem. The resulting protocol is very simple and consists of the following two points. First, all users publish their encryption key. Secondly, messages are sent to each user A encrypted by A's encryption key.

Protocols are designed with a specific aim in mind. Both the security properties of the underlying cryptosystem and those of the protocol itself have to be taken into consideration in the evaluation of the protocol. For instance, in regard to the very simple protocol described above, the underlying cryptosystem might be safe, but the protocol as such still does not prevent the possibility of impersonating: Some user C might pretend to be the user B when sending messages to A. To prevent the occurrence of such situations, some convention of signing messages has to be added to the protocol.

The purpose of this section is to give an overview of some of the most widely studied protocols. We begin with a protocol for playing poker by telephone. This is perhaps the oldest of the "impossible" problems in communication approached from the point of view of cryptography.

Example 7.12. Two persons A and B want to play poker by telephone without any third party acting as an impartial judge. We consider the basic variant of the game, where five cards are dealt. The problem is to deal cards in such a way that the following conditions hold:

(i) All hands (sets of five cards) are equally likely.
(ii) The hands of A and B are disjoint.
(iii) Both players know their own hands but have no information about the opponent's hand.
(iv) It is eventually possible for each of the players to find out the cheating of the other player.

To satisfy (i)–(iv), it is obviously necessary for A and B to exchange information in encrypted form. In fact, several padlocks (in the sense explained in Section 7.2) might have to be used. Therefore, it is desirable that the underlying cryptosystem is commutative. This holds true with respect to the cryptosystem used in the protocol described below. The system is related to the RSA system and is based on the fact that it is difficult to compute *discrete logarithms*—that is, to find an exponent x such that

$$a^x \equiv b \,(\mathrm{mod}\,p),$$

where the integers a and b and the prime p are given.

Players A and B first agree on a large prime p, as well as on the representation of the 52 cards by distinct integers $w_1, \ldots, w_{52}$ in the interval $[2, p-1]$. Then A chooses (and keeps secret from B) two integers e_A and d_A such that

$$e_A d_A \equiv 1 \,(\mathrm{mod}\,p-1).$$

The integers e_A and d_A are referred to as *encryption* and *decryption exponents*. The encryption of a number w is carried out by taking the smallest positive remainder $e_A(w)$ of $w^{e_A} \,(\mathrm{mod}\,p)$. Clearly,

$$(w^{e_A})^{d_A} \equiv w \,(\mathrm{mod}\,p).$$

Similarly, player B chooses (and keeps secret) two integers e_B and d_B such that

$$e_B d_B \equiv 1 \,(\mathrm{mod}\,p-1).$$

Observe that the underlying cryptosystem is commutative: the order of encryptions and decryptions does not matter.

The choices mentioned above can be viewed as Step 0 in the protocol whose other four steps will now be presented. In the protocol A acts as the dealer, but, of course, the roles of A and B can be interchanged if several hands are dealt.

Step 1. B shuffles the cards, encrypts them using the encryption exponent e_B, and tells the result to A. In other words, B tells A the elements of the sequence

$$e_B(w_1), \ldots, e_B(w_{52}) \qquad (24)$$

in a randomly chosen order.

Step 2. A chooses five of the numbers (24) and tells them to B. These numbers represent B's cards.

Step 3. A chooses another five of the numbers from (24), encrypts them by e_A, and tells the resulting five numbers to B.

Step 4. B uses d_B to decrypt the five numbers from Step 3 and tells the result to A. These five numbers represent A's cards.

If this protocol is followed, it is clear that the two hands will be disjoint and that both players know their own cards. In particular, because

$$d_A(d_B(e_A(e_B(w_i)))) = d_A(e_A(d_B(e_B(w_i)))) = w_i,$$

player A knows his or her cards after Step 4.

But does this protocol satisfy all conditions (i)–(iv)? At a first glance, it does. When the cards have the padlock e_B, A does not know them. Similarly, B does not know the cards locked with e_A. However, there are ways of obtaining useful *partial* information. If A makes use of such information, then he or she might want to choose certain types of hands for B and certain other types of hands for himself or herself. We mention one example.

A number c is called a **quadratic residue** (mod n) if the congruence (23), where $e = 2$, has a solution. Otherwise, c is a **quadratic nonresidue** (mod n).

In our case the modulus is a prime number p. In this case it can quickly be computed whether or not a given number is a quadratic residue. (See the survey of results at the end of Section 7.4.) It is also clear that if c is a quadratic residue (resp. nonresidue), then so is the encrypted version of c, obtained by using some encryption exponent.

Assume now that A realizes that in the representation $w_1, \ldots, w_{52}$, all aces are quadratic residues. Hence, in the representation of (24), all aces are also quadratic residues. In this case it is clear that A should give only quadratic nonresidues in Step 2 and only quadratic residues in Step 3. After this, A knows that B has no aces. It is also clear that all hands are not equally likely.

The protocol can be protected against this loophole simply by choosing all the numbers $w_1, \ldots, w_{52}$ to be quadratic residues. However, a similar loophole is obtained by considering cubic residues or arbitrary power residues (that is, in (23) e is arbitrary). The reader is encouraged to consider the security of the protocol against similar attacks based on arithmetic, still assuming that the underlying cryptosystem is secure—that is, the problem of finding discrete logarithms is intractable.

There are some more-sophisticated (and complicated!) protocols such that one can actually prove that it is impossible for the players to obtain partial information about the opponent's cards. Of course, such proofs have been based on the assumption that some specific problem is intractable. For the protocol given in [GM], this problem is factorization: If one is able to make an educated guess about a specific bit in some of the opponent's cards (more specifically, if the guess is correct in 51% of the cases), then one is also able to factor the product of two large primes in random polynomial time. Hence, assuming that the latter problem is intractable, an educated guess is also not possible.

Observe, finally, that at the end all secret information (such as the encryption and decryption exponents in the protocol above) should be made available to the opponent in order to facilitate the disclosure of eventual cheating.   □

We now consider the general problem of how two parties, in most cases adversaries, using a communication network or an information system, can generate a *random* sequence of bits. Such a random sequence is required in many protocol designs. Random sequences thus generated are an essential tool in many widely studied protocols, such as those for certified mail and exchange of secrets.

The generation of a random sequence boils down to the generation of one random bit. The basic model for this is flipping a coin. But how can this be accomplished if the parties involved are far apart? The goal we want to achieve is described in the following model of *flipping a coin in a well*.

Two persons, A and B, stand far apart from each other. B is standing by a deep well with very clear water. When A tosses a coin into the well (it is hoped that A does not miss it!), then B knows the outcome of the flip but is unable to change it. On the other hand, A does not know the outcome. Thus, B just tells A what the outcome is. Later on, A may come and look into the well to check whether or not the information given by B was correct.

Basically it does not matter whether A or B flips the coin into the well. However, the model described above is intended to capture the essence of *flipping a coin by telephone*. This is hardly feasible if only one of the parties is active.

So, suppose A and B want to flip a coin by telephone. (A well-known illustration of the situation, due to Manuel Blum, is the following. Alice and Bob have just divorced, live in different cities, and want to decide who gets the car.) The purpose of the next example is to discuss how this is possible.

Example 7.13. A and B are far apart, on different continents. They want to flip a coin by telephone, without any assistance from a third party. More specifically, A should flip a coin to B. At first only B knows the outcome, and B tells it to A. (The outcome might determine the instructions for a certain part of a protocol.) However, it should be possible for A to make sure later on that the information given by B about the outcome was correct.

The following protocol use—in particular, that the congruence (22) can be solved iff n can be factorized—results from Example 7.11.

Step 1. A chooses two large primes p and q and tells B their product $n = pq$.

Step 2. B randomly chooses a number u from the interval $[2, n/2]$ and tells A its square $u^2 = z$. (The computations are carried out modulo n.)

Step 3. A computes $\pm x$ and $\pm y$, the four square roots of z (mod n). This is possible because A knows the factorization of n. Let x' (resp. y') be the smaller of the numbers x and $-x$ (resp. y and $-y$) modulo n.

Step 4. A guesses whether $u = x'$ or $u = y'$. (Observe that because $u < n/2$, $-u$ is greater than u modulo n.) More specifically, A finds the smallest number i such that the ith bit of x' differs from the ith bit of y' and tells B one of the two guesses "the ith bit in your number u is 0" or "the ith bit in your number u is 1."

Step 5. B tells A whether the guess was correct (heads) or wrong (tails).

Step 6. (Later, after an eventual application of some other protocols) B lets A "come near the well" by telling the number u.

Step 7. A releases the factorization of n.

Clearly, A has no way of knowing u and, hence, the guess is a real one. If B could cheat by changing the number u after A's guess (Step 4), this would mean that he or she is able to compute square roots modulo n. Consequently, as shown in Example 7.11, B would be able to factor n. $\square$

Our last examples belong to a rapidly developing area. The area is of special interest because of its wide range of applications. The area consists of problems of the following type. Two or more parties are in the possession of secrets. They want to share some information but not too much. A protocol has to be designed for achieving this goal.

Example 7.14. Two people, A and B, want to find out who is older without learning anything else about each other's age. How can they carry out a conversation satisfying this requirement?

Let us be more specific. We want to design a protocol for the following conversation. At the beginning A knows the integer i and B knows the integer j, namely, the integers indicating A's and B's ages in years. At the end of the conversation, both A and B know whether $i \geq j$ or $i < j$, but A and B have obtained no further information about j and i, respectively.

We assume that the ages are between 21 and 100 years. People whose ages lie outside this range are not likely to be interested in the problem! For technical reasons, we assume that $1 \leq i, j \leq 80$. Thus, $i + 20$ is the age of A in years.

The problem we are considering is usually stated in the following form. Two millionaires want to know who is richer without obtaining any additional information about each other's wealth.

The following protocol is based on a public key cryptosystem. A has published his or her encryption key, E_A. It does not matter which system is used, provided the system satisfies some reasonable assumptions. Of course, E_A should not give away the decryption key, D_A. It is also supposed that $E_A(x)$ assumes "reasonably random" values for randomly chosen values of x.

Step 1. B chooses a large random number x (the number of bits in x has been fixed in advance) and privately computes the value $E_A(x) = k$.

Step 2. B tells A the number $k - j$.

Step 3. A privately computes the numbers

$$y_u = D_A(k - j + u) \text{ for } u = 1, \ldots, 80.$$

Then A chooses a random prime p. (The number of bits in p has been agreed upon in advance. It is somewhat smaller than the number of bits in x.) A computes, for $u = 1, \ldots, 80$, the smallest positive remainder z_u of $y_u \pmod{p}$. It is assumed that, for any two of the numbers z_u, the absolute value of their difference is greater than or equal to 2 and that, moreover, every $z_u < p$. If this condition is not satisfied for the p chosen, another p is tried.

Step 4. A tells B the sequence of numbers (in this order)

$$z_1, \ldots, z_i, z_{i+1} + 1, z_{i+2} + 1, \ldots, z_{80} + 1, p. \tag{25}$$

Step 5. B checks whether or not the jth number in the sequence is congruent to $x \pmod{p}$. If it is, B concludes that $i \geq j$. If it is not, B concludes that $i < j$.

Step 6. B tells A the conclusion.

It is clear that the conclusion made in Step 5 is the correct one. The reader might want to discuss the security of the protocol in view of the fact that the only information from B to A (resp. from A to B) is given in Step 2 (resp. Step 4). Observe also that the scrambling of the numbers $y_u \pmod{p}$ is necessary because of the following reason. If A would simply tell B the sequence

$$y_1, \ldots, y_i, y_{i+1} + 1, y_{i+2} + 1, \ldots, y_{80} + 1,$$

then B could find i by applying E_A to this sequence. The condition of the difference being greater than or equal to 2 guarantees that x is not among the numbers (25) if $i < j$. Thus, B cannot cheat by referring to a false position in the sequence if an eventual checking is carried out. This is especially useful if some safety measures to prevent cheating are added to the protocol. As such, the protocol does not guarantee that B tells the conclusion correctly in Step 6! A more-sophisticated protocol can be designed in which the probability for such cheating becomes vanishingly small.

Example 7.15. The problem considered in the previous example is a special case of the following general problem. The parties $A_1, \ldots, A_n$ each know the definition of a function $f(x_1, \ldots, x_n)$. For simplicity, we may assume that the variables of f range over a finite set and also that the values of f belong to this finite set. Specific values a_i, $i = 1, \ldots, n$, are considered. For each $i = 1, \ldots, n$, A_i knows the value a_i but has no information in regard to the values a_j for $j \neq i$. The parties $A_1, \ldots, A_n$ want to compute the function value $f(a_1, \ldots, a_n)$ without giving away any additional information about the value a_i of their own variables. In other words, a protocol has to be designed such that, after running through the protocol, all parties A_i know the function value $f(a_1, \ldots, a_n)$, but no party A_i has given away any additional information about the value a_i. Here *additional* refers to any information not obtainable from the function value $f(a_1, \ldots, a_n)$.

Thus, in Example 7.14, we were dealing with a function $f(x_1, x_2)$, for $1 \leq x_1, x_2 \leq 80$, such that $f(x_1, x_2) = 1$ if $x_1 \geq x_2$ and $f(x_1, x_2) = 2$ if $x_1 < x_2$.

A protocol—of course, much more complicated than the one in Example 7.14—can be designed for the general problem. However, the security issues are difficult to formalize in the general case: in particular, the issue of collective cheating, where parties form a coalition to cheat the others.

On the other hand, such protocols seem to open really new vistas for confidential communication. For instance, new types of secret votings can be carried out. Some members of a council might have the right of veto. With the new protocols, nobody knows whether a negative decision is based on the majority, or somebody using his or her veto-right, or on both!

Let us consider a specific example. The parties $A, B, C_1, \ldots, C_n$ want to make a yes or no decision. All parties can vote yes or no. In addition, A has a third vote, yes yes, and B has a third vote, no no. It has been agreed in advance that the majority decides if no double votes are cast. (We assume n is odd). In case of a single double vote, the double vote decides. In case of two double votes, the majority of the C_i-votes decides. Such a voting may be visualized as arising in the United Nations, with A and B being superpowers.

Observe that this is again a special case of the general problem mentioned at the beginning of Example 7.15. Hence, if the protocol is followed and A has cast the vote yes, the decision still being no, A does not know whether the decision is due to the majority or to the no no vote from B. On the other hand, if A has cast the vote yes yes and the decision is still no, A

knows that B has cast the vote no no and that the majority of the C's voted no.   □

The area we have discussed in this section is dynamically propagating. We have tried to give a glimpse of some of the most typical phenomena. Among the related topics not considered, we want to mention the problem of exchanging secrets.

In many protocols it is required that at the end both parties disclose their secrets. For instance, the secrets for two parties A_1 and A_2 might consist of two integers n_1 and n_2, both products of two large primes. But suppose the parties do not trust each other and (for instance, being superpowers) do not trust any third party either. Then the exchange of n_1 and n_2 amounts to a nontrivial problem. However, a bit-by-bit exchange protocol, where at each stage one party can find out whether the other is cheating, has been given in [Bl3].

EXERCISES

7.1.   A requirement considered quite important in classical cryptography is that the cryptotext also be in meaningful language. One way to achieve this goal is to add "junk" letters to the cryptotext. For instance, only the first, fourth, seventh, tenth,... letters in the cryptotext might be meaningful. Encrypt the plaintext HAND in this fashion, using CAESAR with the key 4.

7.2.   Discuss the RSA system in terms of the trapdoor principles (i)–(iv) described in Section 7.2.

7.3.   Show by an example that the composition of two trapdoor knapsacks does not necessarily constitute a trapdoor knapsack—that is, a system obtainable by a single multiplier and modulus.

7.4.   Consider the RSA system with $p = 5$ and $q = 7$. Show that the encryption exponent e must always equal the decryption exponent d.

7.5.   Give a simple condition for the morphisms h_0 and h_1 in Example 7.7 such that the sequence of morphisms applied can immediately be recovered from the final word. (*Hint:* Some special letter develops, indicating the sequence of morphisms applied.)

7.6.   In addition to the super-increasing knapsack, give examples of other easy knapsack problems and study their properties as a basis of a cryptosystem.

7.7.   Estimate the complexity of various parts of Shamir's algorithm for breaking the trapdoor knapsack system.

7.8.   Show how to solve fast the congruence $x^2 \equiv a \pmod{n}$, where n is the product of two prime numbers and the factors of n are known. (*Hint:*

First consider the congruence $x^2 \equiv a$ (mod p) where p is prime, and apply the Chinese remainder theorem.)

7.9.   (For those familiar with the reciprocity laws in number theory.) Establish a fast algorithm for computing the Jacobi symbol (a/n) without assuming that the factorization of n is known. Design another protocol for coin flipping, based on quadratic residues and the properties of the Jacobi symbol.

7.10.   Design a protocol for the voting discussed at the end of Example 7.15.

CHAPTER 8

Trends in Automata and Language Theory

8.1. PETRI NETS

The purpose of this chapter is to give an overview of some currently active topics in automata and language theory. The overview is by no means intended to be exhaustive: Some topics have been entirely omitted, and the material within the topics presented has been chosen to give only a general idea of most representative notions and results. As the title of this chapter indicates, the attention is restricted to topics in automata and language theory.

The style of presentation in this chapter is somewhat different from that used in the previous chapters. Most of the proofs are either omitted or only outlined. Sometimes notions are introduced in a not entirely rigorous manner, and results are presented in a descriptive way rather than in the form of precise mathematical statements. We begin with a discussion of *Petri nets*.

In a customary model for computing, the notion of a *state* is quite essential. This is certainly true of most of the models discussed earlier. The notion of a state introduces, at least implicitly, a specific discrete linear time scale for all considerations involved: Time instants can be identified with the current states. Such a linear time scale is not desirable in all considerations. For instance, we might want to model systems where many processors operate independently and in parallel and where some partial computations depend (perhaps in a complicated way) on the outcome of some other computations. In such cases, it is not desirable to specify in detail the order of the various partial processes.

Many phenomena involving concurrency or parallelism are conveniently modeled by Petri nets.

By definition, a **Petri net** is a finite directed graph PN with two types of nodes, referred to as **places** and **transitions.** Every arrow in a Petri net goes either from a place to a transition or from a transition to a place. Consider a transition a. Every place p (resp. q) such that there is an arrow from p to a (resp. from a to q) is referred to as an **input** (resp. **output**) **place** of a. The same place can be both an input and an output place of a.

A **marking** of a Petri net is a mapping m of the set of places into the set of nonnegative integers. The fact that $m(p) = x$ is usually visualized by saying that there are x **tokens** in the place p. A specific **initial marking** m_0 is usually given in the definition of a Petri net.

Thus, formally, a (marked) Petri net is a quadruple PN $= (P, T, A, m_0)$, where the different items are as follows: P and T are nonempty finite disjoint sets (places and transitions). A (arrows) is a subset of $P \times T \cup T \times P$. (It is often also required that the union of the domain and codomain of A equals $P \cup T$—that is, every place and transition is either the beginning or the end of some arrow.) Finally, m_0 (initial marking) is a mapping of P into the set of nonnegative integers.

We now define the **operation,** or **execution,** of a Petri net. A transition is **enabled** (at a marking) iff all of its input places have at least one token. An enabled transition may **fire** by removing one token from each of its input places and adding one token to each of its output places.

The operation of a Petri net starts with the initial marking m_0. Whenever some transition is enabled, it may fire. This leads to a new marking. If more than one transition is enabled, the firing of such transitions is viewed in an asynchronous fashion: They may fire simultaneously or at different times, one after the other. If two transitions have common input places, they are said to be in **conflict.** This means that only one of them can fire at any marking.

The reader is encouraged to formalize the given definitions—in particular, the change in the marking caused by firing. It is obvious that the formalization leads to a vector addition system in the sense of Section 5.3.

Example 8.1.   We begin with some rather simple Petri nets. In general, when a Petri net is used to model a system, then transitions can be visualized to represent events in the modeled system, while places represent conditions or resources.

In the subsequent figures, places are drawn as circles and transitions as bars. Marking is indicated by the appropriate number of dots in a place.

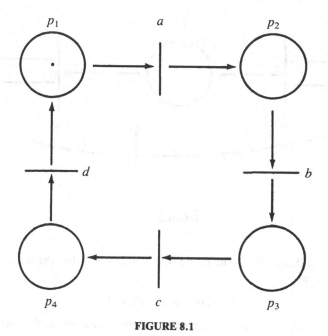

FIGURE 8.1

Figure 8.1 shows our first Petri net, PN_1. The figure gives also the initial marking. Whenever the places are numbered, markings can be defined by vectors in the natural way. Our figure indicates a linear order of the places by the labels p_i. Consequently, the initial marking can be defined by the vector $(1, 0, 0, 0)$.

Initially, only the transition a is enabled. After a has fired, the marking is $(0, 1, 0, 0)$ and only the transition b is enabled. At any moment of time, exactly one transition of PN_1 is enabled. Consequently, conflicts never arise. The markings

$$(1, 0, 0, 0), \qquad (0, 1, 0, 0), \qquad (0, 0, 1, 0), \qquad (0, 0, 0, 1)$$

are the only possible ones for PN_1.

The Petri net PN_1 depicts a system where four stages alternate in a cyclic fashion. For instance, p_1–p_4 could be visualized as the four seasons winter, spring, summer, and fall. Then a is the transition from winter to spring, and so forth.

Consider the Petri net PN_2 in Figure 8.2. The initial marking $(1, 0, 0)$ is also indicated in the picture. Initially, transitions a and c are in conflict and continue to be so until transition c has fired. After that only transition b can be enabled. It continues to be so until the place p_2 runs out of the to-

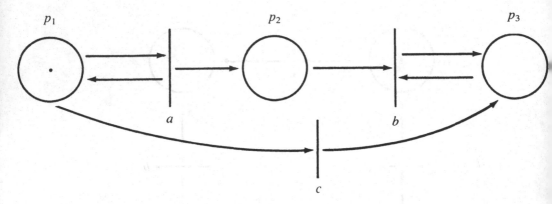

p_1 p_2 p_3

a b

c

FIGURE 8.2

kens gathered there before the firing of c. This means that the markings possible for PN_2 are

$$(1, k, 0) \quad \text{and} \quad (0, k, 1)$$

where k runs through all nonnegative integers. Moreover, for each k, the marking $(0, k, 1)$ is possible only if the marking $(1, k, 0)$ has occurred earlier. In contrast with PN_1, infinitely many markings are possible for PN_2.

Our third Petri net, PN_3, is defined by Figure 8.3. Here the initial marking is $(0, 0, 0)$. Observe that a and b are always enabled because their input condition is satisfied vacuously. Arbitrarily many tokens can be collected in p_1 and/or p_2 by firings of a and/or b. By firing c, the tokens can be passed on to p_3. Every marking (x, y, z) is possible for PN_3, the only exception being that markings $(0, y, z)$ with $z \geq 1$ are not possible.

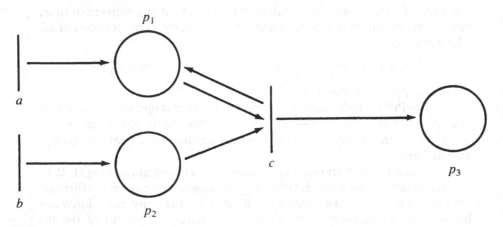

p_1

a

c

p_3

b

p_2

FIGURE 8.3

Petri net PN_3 depicts the basic mode of operation in a public key cryptosystem. Transition a corresponds to creating a public key for a particular user A. Transition b stands for compiling a plaintext. Transition c corresponds to sending an encrypted message to A. This can be done if both some plaintext and public key are available. The latter continues to be available, and it does not matter if several public keys are created.

We conclude Example 8.1 by pointing out that a number of matters discussed in this book can be represented in terms of Petri nets. A typical example is the diagram in Section 3.2 corresponding to Theorem 3.3. The places in this diagram are labeled by AFDA, AFNA, DRE, and Regular. The "reasons" for the arrows correspond to transitions. Thus, Theorem 2.1 can be viewed as the transition by which one can go from a regular language to an AFDA language. Markings are analogous to those of PN_1, and the initial position of the token can be chosen arbitrarily.   □

In Example 8.1 we listed all the possible markings for each of the Petri nets discussed. In general, we say that a marking m is **reachable** for a Petri net, PN, iff there is a finite sequence of firings of the transitions of PN leading PN from the initial marking m_0 to m. The **reachability graph** of a Petri net, PN, is a (not necessarily finite) directed graph whose nodes are labeled by the markings reachable for PN. The arrows of the reachability graph are labeled by the transitions of PN in such a way that there is an arrow labeled by a from a node labeled by m_1 to a node labeled by m_2 iff the firing of a changes the marking from m_1 to m_2.

For instance, the reachability graphs of the Petri nets PN_1 and PN_2 discussed in Example 8.1 are shown in Figure 8.4. Thus, the graph of PN_2 is infinite. The reader might want to draw the reachability graph of PN_3 (which is also infinite).

The reachability graph of a Petri net PN is finite iff PN is **place-bounded**: There is a constant k such that $m(p) \leq k$ holds for all reachable markings m and places p. This condition is automatically satisfied for the following modification of Petri nets.

A Petri net with **place capacities** is a pair (PN, C) where PN is a Petri net (as defined above) and C is a mapping of the set of places of PN into the

$$(1,0,0,0) \xrightarrow{a} (0,1,0,0)$$
$$\uparrow d \qquad\qquad \downarrow b$$
$$(0,0,0,1) \xleftarrow{c} (0,0,1,0)$$

$$(1,0,0) \xrightarrow{a} (1,1,0) \xrightarrow{a} \dots \xrightarrow{a} (1,k,0) \xrightarrow{a} (1,k+1,0) \xrightarrow{a} \dots$$
$$\downarrow c \qquad\quad \downarrow c \qquad\qquad\qquad\quad \downarrow c \qquad\qquad\quad \downarrow c$$
$$(0,0,1) \xrightarrow{b} (0,1,1) \xrightarrow{b} \dots \xrightarrow{b} (0,k,1) \xrightarrow{b} (0,k+1,1) \xrightarrow{b} \dots$$

FIGURE 8.4

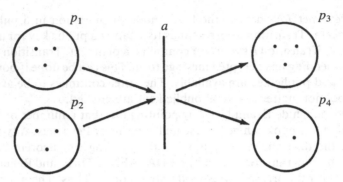

FIGURE 8.5

set of positive integers. For a place p, the value $C(p)$ is referred to as the
capacity of the place p.

Our previous definitions remain unaltered for Petri nets with place
capacities, with the following exception. The number of tokens in a place p
may never exceed $C(p)$; that is, we must have $m(p) \leq C(p)$ for all reachable
markings m. This imposes the following *additional* condition for a transition
a to be enabled: Every output place p of a has at most $C(p) - 1$ tokens. For
instance, the transition a of Figure 8.5 is not enabled if $C(p_3) = 3$ or
$C(p_4) = 2$.

Example 8.2.   Consider the Petri net PN_4 of Figure 8.6, due to [St]. Again,
the initial marking $(1, 1, 1)$ is indicated in the figure. We define further

$$C(p_1) = C(p_2) = C(p_3) = 2,$$

introducing place capacities in this fashion.

The notion of a reachability graph can, of course, be extended to
concern Petri nets with place capacities. The reachability graph of (PN_4, C)
is given in Figure 8.7. The graph shows that after five or seven firings, a
marking $((2, 0, 2)$ or $(2, 1, 2))$ where no firings are possible any more is always

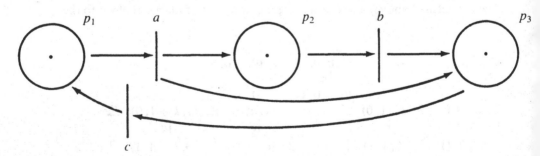

FIGURE 8.6

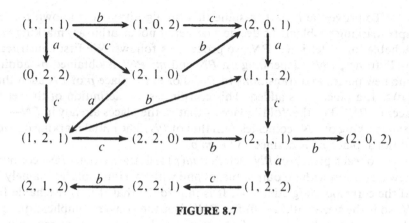

FIGURE 8.7

reached. The sequences of five and seven firings can both be chosen in six different ways.   □

The **reachability problem** for Petri nets consists of finding an algorithm for deciding about a Petri net *PN* and a marking *m* of *PN* whether or not *m* is reachable for *PN*. The reachability problem was for a long time the most-celebrated open problem in the theory of Petri nets.

It is clear that the decidability of the reachability problem for vector-addition systems implies the decidability of the reachability problem for Petri nets. Hence, the following theorem is an immediate consequence of the considerations in Section 5.3.

Theorem 8.1. *The reachability problem for Petri nets is decidable.*

Before a proof for Theorem 8.1 was found, many equivalent versions for the statement of the theorem were known. The equivalent versions dealt with various aspects of Petri nets, language theory, and vector-addition systems. An example is given in Lemma 8.2. The proof of Lemma 8.2 is straightforward and completely independent of the very involved proof of Theorem 8.1.

The **empty marking problem** for Petri nets consists of finding an algorithm for deciding whether or not the marking $(0, \ldots, 0)$ is reachable for a given Petri net.

Lemma 8.2. *The reachability problem for Petri nets is decidable iff the empty marking problem for Petri nets is decidable.*

Proof. The *only if* part is clear since the empty marking problem is a special case of the reachability problem.

To prove the *if* part, assume that an algorithm A is known for the empty marking problem. To decide whether or not an arbitrary marking m is reachable for a Petri net, PN, we proceed as follows. We first construct a new Petri net, PN', depending on PN and m. PN' is obtained by adding some new places and transitions to PN. For every place p of PN such that $m(p) \geq 1$, a place p' is added. This constitutes the definition of the set of places of PN'. For the "old" places—that is, the places already in PN—the initial marking of PN' coincides with that of PN. The initial marking of each additional place p' is defined to be $m(p)$.

For each place p of PN such that $m(p) \geq 1$, the Petri net PN' contains a new transition with no output places and with two input places, namely, p and the corresponding added p'. It is immediate that if m is reachable for PN, so is the vector $(0, \ldots, 0)$ for PN'. But the converse implication also holds true, since the additional places p' can be used only for removing a token simultaneously from p' and the corresponding p. Consequently, an application of the algorithm A to the Petri net PN' settles the question of whether or not m is reachable for PN.   □

A further modification of Petri nets is obtained by introducing **multiplicities** for arrows. (In this way an alternative and a somewhat simpler proof for Lemma 8.2 is also obtained.) Multiplicities mean that to each arrow there is an associated integer greater than or equal to 1. The multiplicity of an arrow indicates the number of tokens to be subtracted from an input place, as well as the number of tokens to be added to an output place. A transition is not enabled if there are not sufficiently many tokens in each of its input places or if the capacity of some of its output places will be exceeded. (The latter condition is required only in case we are considering Petri nets with place capacities.)

We say that a Petri net PN is **alive** iff whenever m is a reachable marking for PN and a is a transition of PN, it is possible for PN to reach, starting from m, a marking in which a is enabled.

Thus, intuitively, a Petri net being alive means that no matter how random sequence of firings we have chosen, it is still possible after this sequence of firings to reach a marking where our favorite transition is enabled!

It is a fairly easy consequence of Theorem 8.1 that the problem of whether or not a given Petri net is alive is decidable. From the point of view of complexity theory, both this problem and the reachability problem seem to be intractable.

We finally discuss **languages** defined by Petri nets. Provided each transition has a distinct label (as is the case in most of the examples considered above), sequences of firings can be represented by words over the label alphabet. The **exhaustive language** $L(PN)$ of a Petri net PN consists of all words leading from the initial marking to some reachable marking. The exhaustive language of the Petri net PN_1 considered in Example 8.1 is denoted by the regular expression

$$(abcd)^*(\lambda \cup a \cup ab \cup abc).$$

On the other hand, we have

$$L(PN_2) = \{a^i \mid i \geq 0\} \cup \{a^i cb^j \mid i \geq j \geq 0\}.$$

The language $L(PN_3)$ consists of all words w over the alphabet $\{a, b, c\}$ such that every prefix of w contains at least as many b's as c's and the first occurrence of c (if any) is preceded by at least one occurrence of a. Thus, $bbabcbc$ is $L(PN_3)$, whereas $bbcabb$ and $abcc$ are not in $L(PN_3)$. Finally, we see that $L(PN_4)$ consists of all prefixes of the language

$$((bc \cup cb)ac \cup (ac \cup ca)(bc \cup cb))(b \cup acb).$$

Consequently, $L(PN_1)$ and $L(PN_4)$ are regular languages, whereas $L(PN_2)$ and $L(PN_3)$ are context-free nonregular languages.

Exhaustive languages of Petri nets have the special property of being prefix-closed: Every prefix of a word in the language is also in the language. The following definition provides more leeway in this respect.

Let PN be a Petri net and M a finite set of reachable markings for PN. Then the language $L(PN, M)$ of PN *with respect to the final markings in M* consists of all words (over the alphabet of labels for transitions) leading from the initial marking to some marking in M.

For instance, we obtain

$$L(PN_2, \{(0, 0, 1)\}) = \{a^i cb^i \mid i \geq 0\}.$$

We shall not go into a detailed study of the language family consisting of languages $L(PN, M)$. Further variations of these families are obtained by considering the modifications of Petri nets mentioned above: nets with place capacities and arrow multiplicities. In fact, the language $L(PN_4)$ discussed above concerns such a modification.

Example 8.3. Languages of the form $L(PN, M)$ are not necessarily context-free. For instance, consider the very simple Petri net PN_5 in Figure 8.8. Thus,

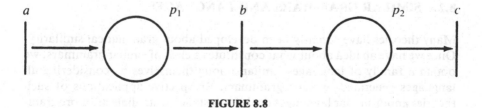

FIGURE 8.8

the initial marking is $(0, 0)$. Define $M = \{(0, 0)\}$. We claim that the language $L = L(PN_5, M)$ is not context-free. Indeed, L consists of all words w over the alphabet $\{a, b, c\}$ such that w contains the same number of a's, b's, and c's and, moreover, there is no prefix of w where the number of b's exceeds that of a's or the number of c's exceeds that of b's. It is easy to see (for instance, by Theorem 3.13) that L is not context-free.

For further examples of non-context-free languages of the form $L(PN, M)$, see Exercise 8.1.

On the other hand, not even all finite languages are of the form $L(PN, M)$. Examples of this are the languages

$$\{aab, baa\} \quad \text{and} \quad \{aba^2ba^3ba^4b\}.$$

The reader is referred to [St] for further details.   □

Petri net languages resemble languages of basic L systems in the sense that some simple languages, even finite ones, are missed. This disadvantage can be removed by considering auxiliary operations. The following theorem, whose proof is left to the reader, provides an example.

Theorem 8.3. *Every regular language is obtained by a letter-to-letter morphism from a language of the form $L(PN, M)$.*

The Petri net languages discussed above deal with sequentialized versions of Petri nets: Words correspond to sequences of firings. On the other hand, Petri nets model systems exhibiting concurrent behavior. Concurrency in the model means simultaneous firing of several transitions. In this way a firing sequence will be a sequence of subsets of the set of all transitions. The **subset language** of a Petri net consists of all such subset firing sequences. (Here again, both the exhaustive language and the language with respect to some set of final markings may be considered.) This approach for defining the language seems more appropriate for Petri nets. One can also show that there exist nets that are equivalent from the point of view of string languages but nonequivalent from the point of view of subset languages.

8.2. SIMILAR GRAMMARS AND LANGUAGES

Many theories have recently been developed about grammatical similarity. Once we have an idea about what constitutes a class of similar grammars, we obtain a family of languages similar among themselves by considering all languages generated by such grammars. Prospective applications of such theories about similar languages deal, for instance, with dialects of programming languages or (in regard to L systems) with species of organisms.

The purpose of this section is not to give an overview about various theories concerning grammatical similarity. (The reader is referred to [Wo] for such an overview.) We restrict our attention to language families generated by grammars. Customarily this theory is referred to as the theory of **grammar forms.**

Apart from giving insight about grammatical similarity, the theory of grammar forms sheds light on various aspects concerning the grammars and languages involved. Here the attention is focused on *context-free* grammars. We shall discuss neither grammars more general than context-free grammars nor other types of generative devices, such as L systems. For extensions and generalizations in these directions, the reader is referred to [Wo].

We now begin the formal details of our discussion. Consider some production in a context-free grammar, for instance,

$$A \rightarrow aBbBAa. \tag{1}$$

Production (1) indicates a *pattern* of terminals and nonterminals. The pattern is (on the right side): T-NT-T-NT-NT-T. Moreover, (1) tells us, for instance, that the first two terminals are different, and so are the last two nonterminals. On the other hand, for instance, the first and the last terminals may either coincide or be different—they are just of the form a.

Productions following this pattern are viewed as **similar** to production (1). This is the idea behind the following definitions.

Context-free grammars will be written as

$$G = (\Sigma_N, \Sigma_T, P, S),$$

where Σ_N and Σ_T are the (disjoint) alphabets of nonterminals and terminals, P is the set of (context-free) productions and S is the start letter. Let h be a letter-to-letter morphism mapping both $\Sigma_N{}^*$ into $\Sigma_N{}'^*$ and $\Sigma_T{}^*$ into $\Sigma_T{}'^*$, for some disjoint alphabets $\Sigma_N{}'$ and $\Sigma_T{}'$. Assume, further, that every letter of $\Sigma_N{}' \cup \Sigma_T{}'$ appears as an h-image. Thus, the cardinality of $\Sigma_N{}'$ (resp. $\Sigma_T{}'$) equals at most that of Σ_N (resp. Σ_T), and we may write

$$h(\Sigma_N) = \Sigma_N{}' \quad \text{and} \quad h(\Sigma_T) = \Sigma_T{}'.$$

The set $h(P)$ is defined in the natural way by

$$h(P) = \{h(\alpha) \rightarrow h(\beta) \mid \alpha \rightarrow \beta \text{ in } P\}.$$

Observe now that $G' = (\Sigma_N{}', \Sigma_T{}', h(P), h(S))$ is also a context-free grammar. We call G' a **morphic image** of G and let

$$G' = h(G). \tag{2}$$

The requirement of h being letter-to-letter is essential because, otherwise, we do not even obtain proper alphabets for the new grammar, G'.

A grammar G_1 is a *subgrammar* of a grammar G_2 iff each of the first three items of G_1 is a subset of the corresponding item of G_2, and the start letters of G_1 and G_2 coincide. (As already pointed out, all grammars in this section will be context-free even if this is not explicitly mentioned.)

A grammar G is said to be an **inverse morphic image** of a grammar G'' iff (2) is satisfied for some letter-to-letter morphism h and a subgrammar G' of G''. The fact that G is an inverse morphic image of G'' under a letter-to-letter morphism h is denoted by

$$G \in h^{-1}(G'').   \tag{3}$$

The reader might wonder why (2) and (3) are not symmetric. A "supergrammar" G'' of G' is used in (3) to provide more leeway: A grammar is an inverse morphic image of another grammar if the former can be mapped morphically onto a subgrammar of the latter. It is here quite essential that we are able to use only a subset of the production set.

The **language family** generated by a grammar G is defined by

$$\mathcal{L}(G) = \{L(G_1) \mid G_1 \in h^{-1}(G) \text{ for some } h\}.$$

A family $\mathcal{L}$ of languages is called **grammatical** iff $\mathcal{L} = \mathcal{L}(G)$ for some G. Two grammars G_1 and G_2 are called **family equivalent** iff $\mathcal{L}(G_1) = \mathcal{L}(G_2)$. (Recall that G_1 and G_2 being equivalent means that they generate the same language.) It is easy to see that two grammars can be family equivalent without being equivalent, and vice versa. For this it is sufficient to observe that a renaming of the terminal alphabet of a grammar G (such as providing all terminals with primes) leaves the family $\mathcal{L}(G)$ invariant, whereas, in general, it changes $L(G)$.

To avoid unnecessary special cases, we consider only λ-free languages in this section. When speaking of specific language families, such as the families of regular and context-free languages, we mean the collection of λ-free languages in these families. Moreover, it is assumed that the grammars considered do not contain productions $A \to \lambda$.

If we study the family $\mathcal{L}(G)$ associated to a grammar G, then G itself is often referred to as a **grammar form**. Thus, grammars and grammar forms coincide as objects, but only $L(G)$ is studied in connection with a grammar.

In the terminology dealing with grammar forms, inverse morphic images are often called **interpretations**. The inverse morphism can be replaced by a direct operation as follows. Consider a grammar, $G = (\Sigma_N, \Sigma_T, P, S)$, viewed here as a grammar form. Let σ be a finite substitution satisfying the following conditions. For each letter a in $\Sigma_N \cup \Sigma_T$, $\sigma(a)$ is a finite set of letters such that $\sigma(a) \cap \sigma(b)$ is empty whenever $a \neq b$. The set $\sigma(P)$ consists of all productions obtained from some production in P by replacing every letter a with some letter in $\sigma(a)$.

For instance, if P consists of the single production $S \to aSA$ and $\sigma(S) = \{S\}$, $\sigma(A) = \{A_1, A_2, A_3\}$, and $\sigma(a) = \{a_1, a_2\}$, then

$$\sigma(P) = \{S \to a_1 SA_1, S \to a_1 SA_2, S \to a_1 SA_3, S \to a_2 SA_1, S \to a_2 SA_2, S \to a_2 SA_3\}.$$

Here the σ-images are obtained by providing the original letters with indices. It is always useful to visualize σ-images in this fashion.

Let us now denote $\sigma(\Sigma_N) = \Sigma_N'$ and $\sigma(\Sigma_T) = \Sigma_T'$. Assume that S' is in $\sigma(S)$ and P' is a subset of $\sigma(P)$. Then the grammar

$$G' = (\Sigma_N', \Sigma_T', P', S')$$

is said to be an **interpretation** of G.

It is immediate that this notion of an interpretation coincides with the above one—that is, with the notion of an inverse morphic image. The set of all interpretations of G is denoted by $\mathfrak{M}(G)$.

The following lemma contains material obtainable directly from the definitions.

Lemma 8.4. *A grammar G_1 is an interpretation of a grammar G_2 iff $\mathfrak{M}(G_1) \subseteq \mathfrak{M}(G_2)$. If G_1 is an interpretation of G_2, then $\mathcal{L}(G_1) \subseteq \mathcal{L}(G_2)$ but not necessarily vice versa. There is an algorithm for deciding of two given grammars G_1 and G_2 whether or not G_1 is an interpretation of G_2 and, hence, also an algorithm for deciding whether or not $\mathfrak{M}(G_1) = \mathfrak{M}(G_2)$. For every language L in $\mathcal{L}(G)$, there is a letter-to-letter morphism h such that $h(L) \subseteq L(G)$. The language $L(G)$ is finite iff every language in $\mathcal{L}(G)$ is finite. If $L(G) = \{w_1, \ldots, w_n\}$, then G is family equivalent to the grammar determined by the productions*

$$S \to w_1, \ldots, S \to w_n.$$

The last sentence of Lemma 8.4 tells us that the *family* $\mathcal{L}(G)$ is completely determined by the *language* $L(G)$, provided the latter is finite. This is definitely not the case for arbitrary grammars G. The following example provides further illustration for this and other phenomena.

Example 8.4. Each of the following grammars G_1–G_7 is defined by listing the productions:

G_1: $S \to a, S \to b, S \to ab, S \to ba$;

G_2: $S \to a, S \to aa$;

G_3: $S \to aS, S \to a$;

G_4: $S \to aSa, S \to a, S \to a^2$;

G_5: $S \to aS, S \to Sa, S \to a$;

G_6: $S \to A, S \to B, A \to A^2, A \to a^2, B \to aB, B \to Ba, B \to a$;

G_7: $S \to SS, S \to a$.

Here only $L(G_1)$ and $L(G_2)$ are finite. Hence, the language families of G_1 and G_2 are completely determined by their languages. Clearly, $\mathcal{L}(G_2)$ consists of all finite languages where every word is of length 1 or 2. On the other hand, every language in $\mathcal{L}(G_1)$ has also this property, but $\mathcal{L}(G_1)$ does not contain all such languages and, consequently, $\mathcal{L}(G_1) \subsetneqq \mathcal{L}(G_2)$. It is obvious that the language $\{aa\}$ is not in $\mathcal{L}(G_1)$. The reader might want to prove that the language $\{ab, bc, ca\}$ is not in $\mathcal{L}(G_1)$. A contradiction is easily derived when considering different possibilities for the definition of σ.

It is also easy to establish the strict inclusion $\mathcal{L}(G_3) \subsetneqq \mathcal{L}(G_5)$. For instance, $\{a^n b^n \mid n \geq 1\}$ is a language in the difference of the two families. Observe, however, that $L(G_3) = L(G_5) = a^+$.

In fact, one can show by the results presented later on in this section that

$$\mathcal{L}(G_i) \subsetneqq \mathcal{L}(G_{i+1}) \qquad \text{for } 1 \leq i \leq 6.$$

It is easy to see by the theory developed in Chapter 3 that $\mathcal{L}(G_3)$ consists of all regular languages. It is a direct consequence of Theorem 2.2 that $\mathcal{L}(G_7)$ equals the family of context-free languages. (Recall our convention: We consider only λ-free languages in this section.) $\square$

A grammar is called **regular-complete** (resp. **regular-sufficient,** iff its language family equals (resp. contains) the family of regular languages. A grammar is called **context-free complete**—or, briefly, **complete**—iff its language family equals the family of context-free languages. (Since we consider only context-free grammars, it does not make sense to talk about context-free sufficiency.)

The notions of completeness and sufficiency can be defined for other language families as well. In particular, we consider the family of linear languages in the sequel. By definition, a context-free grammar is **linear** iff there is at most one nonterminal on the right side of every production. A context-free language is **linear** iff it is generated by some linear grammar.

Every complete grammar G can be viewed as a *normal form* for all context-free grammars: Every context-free language is generated by a grammar whose productions follow the pattern of the productions in G. Consequently, a characterization of complete grammars is at the same time a characterization of all possible normal forms for context-free grammars. Similarly, a characterization of regular-complete (resp. linear-complete) grammars is at the same time a characterization of all possible normal forms for grammars for regular (resp. linear) languages.

A characterization of regular-complete, linear-complete, and complete grammars will be given later. We begin with a notion very useful in this characterization.

It is often very difficult to handle the situation where there are several terminal letters in a grammar, because the interpretations of the terminals have to be disjoint. A grammar is termed **unary** (with respect to terminals) if it has only one terminal letter. Thus, the grammars G_2-G_7 in Example 8.4 are unary.

Whenever a grammar G has several terminal letters, we consider *a*-**restrictions** G_a of G, where a is some terminal letter of G. Such an *a*-restriction is obtained from G by removing all productions containing terminals $b \neq a$.

We say that a grammatical family $\mathcal{L}$ is **unary-complete** iff whenever a grammar G is not unary and satisfies $\mathcal{L}(G) = \mathcal{L}$, then G possesses an *a*-restriction G_a also satisfying $\mathcal{L}(G_a) = \mathcal{L}$.

Thus, if $\mathcal{L}$ is unary-complete and $\mathcal{L}(G) = \mathcal{L}$, where G has several terminals, we know that all the terminals except one are superfluous in the sense that they are not actually needed in the generation of $\mathcal{L}$.

It is clear that all grammatical families are not unary-complete. Considering the grammar G_1 in Example 8.4, it is immediately seen that $\mathcal{L}(G_1)$ is not unary-complete because of the simple reason that $\mathcal{L}(G_1)$ cannot be generated by a unary grammar.

On the other hand, the following theorem shows that the basic families mentioned above are all unary-complete. Apart from the fact concerning regular languages, the proof of the theorem is very involved.

Theorem 8.5. *The families of regular, linear, and context-free languages are all unary-complete.*

No nontrivial examples of families that are not unary-complete are known. More explicitly, no families $\mathcal{L}$ are known such that $\mathcal{L}$ is not unary-complete and $\mathcal{L} = \mathcal{L}(G)$, for some unary grammar G.

By Theorem 8.5, it suffices to consider unary grammars in the characterization of regular-completeness, linear-completeness, and completeness. We denote by a the only terminal letter. If a characterization in the unary case is known, we have to study only *b*-restrictions in the general case for all possible terminal letters b.

We need some simple auxiliary notions. We say that a grammar is **reduced** iff, for every nonterminal $A \neq S$, the start letter S generates a word containing A and, moreover, A generates a word over the terminal alphabet. A grammar G is **self-embedding** iff it has a derivation

$$S \Rightarrow^* w_1 A w_2 \Rightarrow^* w_1 w_3 A w_4 w_2 \Rightarrow^* w_1 w_3 w_5 w_4 w_2,$$

where A is a nonterminal and the w's are terminal words such that $w_3 \neq \lambda$ and $w_4 \neq \lambda$.

We are now able to characterize regular-completeness. All proofs in this section are omitted. However, the proof of the next theorem is not difficult.

Theorem 8.6. *A unary grammar G is regular-complete iff G is not self-embedding and satisfies $L(G) = a^+$. An arbitrary grammar G is regular-*

sufficient iff $L(G) \supseteq a^+$ holds for some terminal letter a. Regular-completeness and regular-sufficiency of an arbitrary grammar are decidable.

Characterization of linear-completeness is much more involved. First, consider grammars G_4 and G_5 in Example 8.4. That G_5 is linear-complete is obvious: Every language in $\mathcal{L}(G_5)$ is linear and, moreover, an arbitrary linear production can be simulated by sufficiently many productions of G_5. For instance, for the simulation of

$$S \rightarrow bAcde$$

the productions

$$S \rightarrow bA_1, \qquad A_1 \rightarrow A_2 e, \qquad A_2 \rightarrow A_3 d, \qquad A_3 \rightarrow Ac$$

can be used. On the other hand, all languages in $\mathcal{L}(G_4)$ are also linear, but G_4 is not linear-complete. It is a consequence of Theorem 8.6 that all regular languages are in $\mathcal{L}(G_4)$. The linear language $\{a^{2i}b^i \mid i \geq 1\}$ is not in $\mathcal{L}(G_4)$. Intuitively, this is due to the restrictions in pumping according to G_4. Only the same number of terminals can be pumped on both sides of the middle.

On the other hand, unrestricted facilities for pumping are necessary for linear-completeness. This idea will now be formalized.

We say that a nonterminal A in a unary reduced grammar G is **left-pumping** (resp. **right-pumping**) iff for some fixed $m, n \geq 0$, there are infinitely many values i such that

$$A \Rightarrow^* a^{i+m} A a^n \qquad (\text{resp. } A \Rightarrow^* a^m A a^{n+i}).$$

The nonterminal A is **pumping** iff it is both left-pumping and right-pumping.

Let $A_1, \ldots, A_m$ be all the pumping nonterminals in a unary reduced grammar G. For each i, the lengths j of the terminal words a^j generated from A_i obviously constitute an ultimately periodic sequence. Denote its period by $p(A_i)$. Let p be the least common multiple of all numbers $p(A_i)$, where $i = 1, \ldots, m$. Denote the residue classes modulo p by $R_0, R_1, \ldots, R_{p-1}$. We say that the residue class R_j is A_i-**reachable** iff there are numbers r, s, and t such that

$$S \Rightarrow^* a^r A_i a^s, A_i \Rightarrow^* a^{t+np} \qquad \text{for all } n \geq 0$$

and

$$j \equiv r + s + t \pmod{p}.$$

The **pumping spectrum** of G consists of all numbers in all A_i-reachable residue classes, where i ranges over $1, \ldots, m$.

It should be emphasized that A_i-reachability as defined above implies that almost all the numbers in the residue class R_j are reachable in the sense defined. The reachability or nonreachability of an "initial mess" does not matter.

Our next theorem characterizes linear-completeness and linear-sufficiency.

Theorem 8.7. *A unary reduced grammar G is linear-sufficient iff* $L(G) = a^+$ *and the pumping spectrum of G consists of all numbers. The same grammar G is linear-complete iff it is linear-sufficient and, in addition, generates no word containing two occurrences of self-embedding nonterminals. Both linear-completeness and linear-sufficiency of an arbitrary grammar are decidable.*

Continuing our previous list of grammars G_1–G_7, we leave it to the reader to prove by Theorem 8.7 that the grammar

$$G_8: \quad S \to a^3 S, S \to Sa^4, S \to a, S \to a^2, S \to a^3$$

is linear-complete, whereas the grammar

$$G_9: \quad S \to A, S \to B, A \to a^2 A, A \to Aa^2, A \to a^2, B \to aB, B \to a$$

is not linear-complete. Since both grammars are linear, they generate only linear languages.

We now turn to the discussion of completeness. The notion corresponding to pumping spectrum is that of an expansion spectrum. We say that a nonterminal A is **expansive** iff

$$A \Rightarrow^* w_1 A w_2 A w_3$$

holds for some terminal words w_1, w_2, and w_3.

Let $A_1, \ldots, A_m$ be all the expansive nonterminals in a unary reduced grammar G. Again, for each i, the lengths j of the terminal words a^j generated by A_i constitute an almost periodic sequence. Denote its period by $p(A_i)$. The least common multiple p and the residue classes R_i are defined exactly as in connection with pumping spectrum. Also, the notion of a residue class R_j being A_i-reachable is defined exactly as before. Finally, the **expansion spectrum** of G consists of all numbers in all A_i-reachable residue classes, where i ranges over $i = 1, \ldots, m$.

Theorem 8.8. *A unary reduced grammar G is complete iff* $L(G) = a^+$ *and the expansion spectrum of G consists of all numbers. The completeness of an arbitrary grammar is a decidable property.*

A major tool in the proof of Theorem 8.8 is the following supernormal-form theorem for context-free languages, a result interesting also on its own right.

Theorem 8.9. *Assume that (i, j, k) is a triple of nonnegative integers. Then every context-free language L is generated by a grammar whose productions are of the two forms*

$$A \to w' \quad and \quad A \to w_i B w_j C w_k,$$

where A, B, and C are (not necessarily distinct) nonterminals and the w's are terminal words such that $|w_i| = i$, $|w_j| = j$, and $|w_k| = k$ and $|w'|$ equals the length of some word in L.

Returning to Example 8.4, it is easy to see by Theorem 8.8 that the grammar G_6 is not complete. Since, clearly, no linear grammar is complete, we conclude that G_7 is the only complete grammar in Example 8.4.

Major open problems concerning grammatical families deal with decidability. For instance, is the family equivalence of two given grammars decidable? It is generally conjectured that this is not the case, at least not for arbitrary context-free grammars. An indication in this direction is that there are "very many" grammatical families. Indeed, the collection of grammatical families possesses a property of density. In general, language families obtained by generative devices very seldom possess such a property.

More specifically, assume that $\mathcal{L}$ and $\mathcal{L}'$ are grammatical families such that $\mathcal{L} \subsetneq \mathcal{L}'$. The pair $(\mathcal{L}, \mathcal{L}')$ is said to be **dense** iff, whenever $\mathcal{L}_1$ and $\mathcal{L}_2$ are grammatical families satisfying

$$\mathcal{L} \subseteq \mathcal{L}_1 \subsetneq \mathcal{L}_2 \subseteq \mathcal{L}',$$

then there is a grammatical family $\mathcal{L}_3$ with the property

$$\mathcal{L}_1 \subsetneq \mathcal{L}_3 \subsetneq \mathcal{L}_2.$$

For instance, if $\mathcal{L}$ and $\mathcal{L}'$ are the families of regular and context-free languages, then one can show that the pair $(\mathcal{L}, \mathcal{L}')$ is dense.

For a long time, it was an open problem whether or not there are two grammars G_1 and G_2 generating finite languages and such that the pair $(\mathcal{L}(G_1), \mathcal{L}(G_2))$ is dense. This problem has now been solved. For instance, the grammars G_1 and G_2 given in Example 8.4 satisfy these conditions.

The ideas presented in this section can also be applied to the *theory of graphs*. This approach seems to be of special interest because of two reasons. In the first place, a classification of graphs is obtained, based on a notion that generalizes the notion of *coloring* in a natural way. In the second place, questions concerning this classification can be identified with questions concerning a fragment of the theory of grammatical families.

Attention is here restricted to (finite) *undirected* graphs. The theory can be carried over to directed graphs as well [MSW4].

For two graphs G and G', we say that G' is **G-colorable** (or **colorable according to** G) iff there is a mapping φ of the set of vertices of G' into the set of vertices of G such that whenever there is a line between two vertices x and y in G', then there is also a line between $\varphi(x)$ and $\varphi(y)$ in G. (We assume that the graphs considered do not contain loops—that is, lines from a vertex to itself—or isolated vertices.) The collection of all G-colorable graphs is denoted by $\mathfrak{M}(G)$. Families $\mathfrak{M}(G)$ are referred to as **color-families.**

The motivation behind this terminology is the following. For $n \geq 2$, denote by K_n the *complete* graph with n vertices—that is, the graph possessing an edge between every two distinct vertices. Clearly, a graph G' is n-colorable in the usual graph-theoretic sense iff G' is in $\mathfrak{M}(K_n)$. Thus, the four-color theorem tells us that every planar graph is in $\mathfrak{M}(K_4)$.

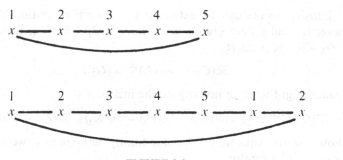

FIGURE 8.9

On the other hand, in the customary theory of colorings, only families $\mathfrak{M}(K_n)$ are considered. The definition above is much more general: colorings according to an arbitrary graph.

For instance, consider the *cyclic* graph C_n with n vertices—that is, the vertices are connected cyclically. Figure 8.9 shows how C_7 is colored according to C_5. Thus, C_7 is in $\mathfrak{M}(C_5)$. It is easy to see that, conversely, C_5 is not in $\mathfrak{M}(C_7)$.

The fact that a graph G' is colorable according to a graph G is analogous to a grammar G' being an interpretation of a grammar G. For instance, the first and third sentences of Lemma 8.4 remain valid for graphs and colorings. The interconnection between graphs and grammars can be stated more explicitly as follows.

Because our graphs are undirected, we have to consider **commutative languages,** which means the order of letters is immaterial. Thus, ab being in the language is equivalent to ba being in the language; this means that the associated graph possesses a line between a and b. For instance, C_5 can be characterized by the commutative language (thus also containing all permutations of the words listed)

$$L_5 = \{a_1a_2, a_2a_3, a_3a_4, a_4a_5, a_5a_1\}.$$

Recall now that according to the last sentence of Lemma 8.4, the family $\mathcal{L}(G)$ is completely determined by the language $L(G)$, provided the latter is finite. Hence, in the finite case, we may speak of interpretations of languages instead of interpretations of grammars. Then the graph C_7 being C_5-colorable means that the commutative language

$$L_7 = \{a_1a_2, a_2a_3, a_3a_4, a_4a_5, a_5a_6, a_6a_7, a_7a_1\}$$

is an interpretation of L_5.

This is true also in general: A graph G' is G-colorable iff the language L' associated with G' is an interpretation of the language L associated with G.

Density results can be extended to concern color-families as well: Whenever G_1 and G_2 are graphs such that $\mathfrak{M}(G_1) \subsetneqq \mathfrak{M}(G_2)$, then there is a graph G_3 with the property

$$\mathfrak{M}(G_1) \subsetneqq \mathfrak{M}(G_3) \subsetneqq \mathfrak{M}(G_2).$$

The reader might want to investigate the inclusions

$$\cdots \mathfrak{M}(C_9) \subsetneqq \mathfrak{M}(C_7) \subsetneqq \mathfrak{M}(C_5) \subsetneqq \mathfrak{M}(C_3) = \mathfrak{M}(K_3) \subsetneqq \mathfrak{M}(K_4) \subsetneqq \mathfrak{M}(K_5) \cdots$$

and show how to "squeeze in" a color-family between any two consecutive families in this hierarchy.

The notion of G-colorability can be extended to concern infinite graphs G'. The notation $\mathfrak{M}_\infty(G)$ stands for all graphs, finite and infinite, colorable according to the finite graph G.

Theorem 8.10. *For all G_1 and G_2, $\mathfrak{M}(G_1) = \mathfrak{M}(G_2)$ iff $\mathfrak{M}_\infty(G_1) = \mathfrak{M}_\infty(G_2)$.*

We discussed above interpretations of finite languages. This notion can be generalized in a natural fashion to concern all languages. Theorem 8.10 can then be extended to this form: Two languages generate the same family of interpretations iff they generate the same family of finite interpretations.

8.3. SYSTOLIC AUTOMATA

Our attention will be restricted in this section to two specific types of **systolic** automata, referred to as systolic **tree** and **trellis** automata. While we do not attempt to define the general notion of a systolic automaton, we begin with an intuitive description of some of its most typical properties.

A systolic automaton receives its input word, say $a_1 \cdots a_n$ where the a's are letters, through a row of input nodes, each of which receives a letter

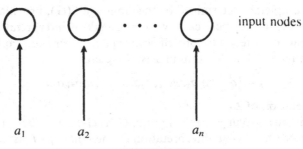

FIGURE 8.10

simultaneously, as in Figure 8.10. The flow of information in the automaton is unidirectional—in our illustrations it is depicted as going up from the bottom.

Mathematically, the automaton is a directed graph. Intuitively, it is viewed as a network of processors: Each node of the graph is a processor that computes a finite function of its inputs and sends the result of the computation as outputs. A typical situation is the one shown in Figure 8.11. After receiving the inputs c_1 and c_2 from some processors "below," the processor A sends the outputs b_1 and b_2 to some processors "above." Formally this means that a function f_A is associated with the node labeled by A. The function f_A maps the set of pairs (c_1, c_2) into the set of pairs (b_1, b_2). Here the c's and b's are letters of a fixed alphabet, referred to as the *operating alphabet* of the automaton.

In this way the information flows "upward" simultaneously all over the network. The network is so constructed that it has a specific topmost node, called the **roof**. The original input $a_1 \cdots a_n$ is accepted iff the output from the roof is an accepting letter: A subset of the operating alphabet, referred to as **accepting letters,** has been specified. All accepted words constitute the **language** accepted by the systolic automaton. The attention is restricted here to systolic automata as acceptors; a translation device is obtained by a suitable modification of the model.

One can visualize a systolic automaton as an abstraction of a VLSI chip. The input is received through a row of input pins. The input is then processed by a specialized, usually large, network of processors on the chip. The unidirectional flow of information guarantees that systolic automata are perfectly suited for "pipelining," an essential requirement for VLSI. Pipelining means essentially that inputs can be pumped in the system at consecu-

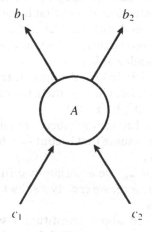

b_1 b_2

A

c_1 c_2

FIGURE 8.11

tive instants of time and, as a result, multiprocessing of the data takes place at each stage of the pipeline. The unidirectional flow of information guarantees that an input will never get mixed up with previous inputs during the processing.

The idea of *pumping* information in the system at consecutive time instants originally suggested the term *systolic*.

So far, we have not said anything about the structure of the network defining a systolic automaton—that is, about the interconnections of the nodes in the graph. It is clear that this structure affects the capabilities of the automaton quite deeply. General questions about how this structure affects the capabilities of the automaton so far remain open. On the other hand, specific structures such as a tree structure and the structure of hexagonal interconnections (later referred to as a *trellis*) are well understood. The term *a systolic tree/trellis automaton* is used in the case where the underlying directed graph is a tree or a trellis, respectively.

Before we enter into the details of the two types of automata, we still want to emphasize that, from the point of view of VLSI, it is quite essential that the interconnections in the network be regular and simple. For instance, a very complicated tree structure does not make sense from the point of view of VLSI design. Another point is that if we allow the processors (the nodes in the network) to be different, then the pattern of the different processors should still be very regular.

In general, apart from opening up a new, interesting area of language theory, the study of systolic automata also might give useful hints for VLSI system design, particularly with regard to the pros and cons of specific geometric structures.

We are now ready for more formal details. We begin with the notion of a *systolic tree automaton,* STA. Consider an infinite binary tree T, as in Figure 8.12. Every node in T has two daughters. There is also a specific node (root, or roof, as in our discussion above) that is not a daughter of any node. We assume further that the nodes of T are labeled by letters from an alphabet Σ_p, called the alphabet of **processors,** in such a way that T has only finitely many different infinite, labeled subtrees.

For $k = 0, 1, 2, \ldots$, LEVEL(k) denotes the set of all nodes of T whose distance from the root equals k. Clearly, there is a natural ordering from left to right of the 2^k elements of LEVEL(k).

Let Σ_t and Σ_O be two further alphabets, the **terminal** and the **operating** alphabet, respectively. We assume that Σ_t contains the letter # and that the set $\Sigma_t' = \Sigma_t - \{\#\}$ is not empty.

To every letter A of Σ_p, we associate two functions $g_A : \Sigma_t \rightarrow \Sigma_O$ and $f_A : \Sigma_O \times \Sigma_O \rightarrow \Sigma_O$. Furthermore, we specify a subset Σ_O' of Σ_O, referred to as the set of **accepting letters.**

The items introduced above constitute a **systolic tree automaton,** STA. Our definition has been simplified from the more-general one of [CuSW] in that we consider only a very special class of underlying trees T.

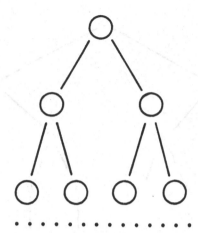

FIGURE 8.12

Given STA, every word w over the alphabet $\Sigma_I{}'$ determines a unique element of Σ_O, denoted by OUTPUT(w), as follows. Assume that w is of length n. Let k be the smallest integer such that LEVEL(k) in T contains at least n nodes. Clearly, k exists and is unique. Let $A_1, \ldots, A_{2^k}$ be the nodes, from left to right, in LEVEL(k). Let $m = 2^k - n$. Then OUTPUT(w, k) is defined to be the word of length 2^k over the alphabet Σ_O whose ith letter, for $i = 1, \ldots, 2^k$, equals the result of applying g_{A_i} to the ith letter of $w\#^m$.

Assume that we have already defined OUTPUT(w, j) $= x_j$ for some j with $1 \leq j \leq k$. Since the length of OUTPUT(w, k) is 2^k, we assume, inductively, that the length of x_j equals 2^j. Hence, we may write

$$x_j = a_1 b_1 a_2 b_2 \cdots a_{2^{j-1}} b_{2^{j-1}},$$

where the a's and b's are letters of Σ_O (possibly identical). Assume further that $B_1, B_2, \ldots, B_{2^{j-1}}$ are the labels of the nodes in LEVEL($j - 1$). Then OUTPUT($w, j - 1$) is defined to be the word of length 2^{j-1} over the alphabet Σ_O whose ith letter, for $i = 1, \ldots, 2^{j-1}$, equals the result of applying f_{B_i} to the argument (a_i, b_i).

Finally, we define

$$\text{OUTPUT}(w) = \text{OUTPUT}(w, 0).$$

Hence, OUTPUT(w) is always a letter of Σ_O. By definition, a word w is **accepted** by STA iff OUTPUT(w) is a letter of $\Sigma_O{}'$.

The **language accepted** by STA consists of all words over the alphabet $\Sigma_I{}'$ accepted by STA.

Example 8.5. Assume that the alphabet of processors, Σ_p, consists of one letter only. Hence, all nodes in the underlying binary infinite tree are labeled by this letter. To define a systolic tree automaton STA in this setup, it suffices to specify the alphabets and two functions, denoted briefly by g and f.

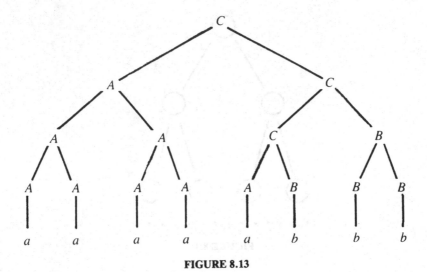

FIGURE 8.13

By definition, the operating alphabet Σ_O of STA equals $\{A, B, C, \#\}$, and the terminal alphabet Σ_t equals $\{a, b, \#\}$. The set Σ_O' of accepting letters consists of C alone. The functions g and f are defined by

$$g(a) = A, \qquad g(b) = B, \qquad g(\#) = \#;$$
$$f(A, A) = A, \quad f(B, B) = B,$$
$$f(A, B) = f(A, C) = f(C, B) = C,$$
$$f(x, y) = \# \qquad \text{in all other cases.}$$

Thus, the input word a^5b^3 gives rise to the computation in Figure 8.13. On the other hand, computations from the inputs a^5b^2 and *abab* are depicted as in Figure 8.14.

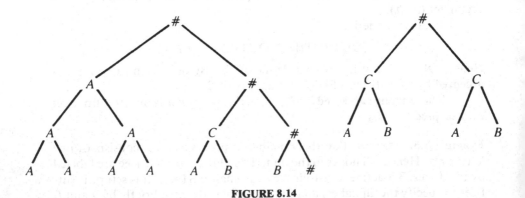

FIGURE 8.14

This means that our STA accepts the word $a^5 b^3$, whereas it rejects the words $a^5 b^2$ and *abab*. It is easy to see that STA accepts the language

$$L(\text{STA}) = \{w \in a^+ b^+ \mid w \text{ is of length } 2^n \text{ for some } n \geq 1\}. \qquad \Box$$

Our first theorem establishes a normal form result useful in many constructions. It says that the node labeling of the underlying tree T is, in fact, unnecessary: We can always obtain the same effect using just one label for the nodes. (The price we pay for the omission of node labels is an increase in the size of the operating alphabet.) Recall that we assumed that T has only finitely many, say n, infinite (labeled) subtrees. Hence, we may number the nodes of T using numbers $1, \ldots, n$ in such a way that if two nodes are roots of identical subtrees, they get the same number. Clearly, if two nodes are numbered by the same number, they are also labeled by the same letter of Σ_p. Moreover, the numbering of the nodes is top-down deterministic: Each number attached to a node uniquely determines the numbers attached to its two daughters. This is the basic idea in the proof of Theorem 8.11. The details of the resulting construction are left to the reader.

We say that a systolic tree automaton is **homogeneous** iff its alphabet of processors consists of one letter only.

Theorem 8.11. *For every systolic tree automaton, an equivalent homogeneous systolic tree automaton can be constructed.*

Theorem 8.12. *The family of languages acceptable by systolic tree automata contains all regular languages and is closed under Boolean operations.*

Proof. Closure under complementation is clear: one only has to replace the set of accepting letters by its complement. Closure under union is established by a straightforward pair construction: the operating alphabet of the new automaton consists of pairs of letters, which makes the simulation of both of the given automata possible. The output at the root is positive iff at least one of the outputs of the given automata is positive. Closure under intersection now follows either from the closure under union and complementation or, directly, by a similar pair construction.

To show that every regular language is accepted by a systolic tree automaton, we first consider an example: the language L denoted by the regular expression $(ab)^*$. L is accepted by the finite deterministic automaton FDA with the following state transition table:

	a	b
q_0	q_1	q_2
q_1	q_2	q_0
q_2	q_2	q_2

Here q_0 is the initial, as well as the only final state.

We now construct a systolic tree automaton, STA, equivalent to FDA. The underlying tree will be in the normal form of Theorem 8.11. The operating alphabet consists of certain sets of pairs whose elements are states of FDA. Which sets are included will be obvious from the following definition of the functions g and f.

By definition,

$$g(a) = \{(q_0, q_1), (q_1, q_2), (q_2, q_2)\} = V_1,$$
$$g(b) = \{(q_0, q_2), (q_1, q_0), (q_2, q_2)\} = V_2,$$
$$g(\#) = \{(q_0, q_0), (q_1, q_1), (q_2, q_2)\} = V_3.$$

Thus, $g(a)$ indicates the state transitions caused by a. It is understood that # causes no changes of states.

The value $f(V_i, V_j)$ is defined to be the product of V_i and V_j when V_i and V_j are viewed as relations. Thus, a pair (q, q') is in $f(V_i, V_j)$ iff there is a q'' such that (q, q'') is in V_i and (q'', q') is in V_j. For instance,

$$f(V_1, V_1) = \{(q_0, q_2), (q_1, q_2), (q_2, q_2)\},$$
$$f(V_1, V_2) = \{(q_0, q_0), (q_1, q_2), (q_2, q_2)\} = V_4,$$
$$f(V_3, V_3) = V_3,$$
$$f(V_4, V_4) = f(V_4, V_3) = V_4.$$

The set of accepting letters consists of those V_i's that contain the pair (q_0, q_0). This completes the definition of STA. The acceptance of the word $(ab)^3$ is illustrated in Figure 8.15.

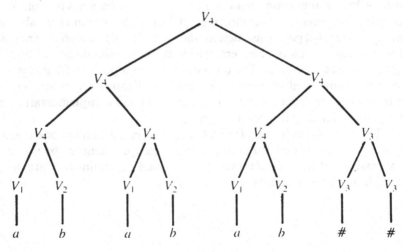

FIGURE 8.15

The reader should have no difficulty in constructing a general proof based on the discussion of the preceding example. In fact, the general definition of g and f is already indicated in connection with the example. Also, in the general case the alphabets are as in the example, with the exception that the set of accepting letters will consist of those V_i's that contain a pair (q_0, q_F), where q_0 is the initial state and q_F is some final state. $\square$

A reader familiar with the basics of algebraic automata theory will notice that the second part of the proof above consists of an STA-simulation of the syntactic monoid defined by the regular language L.

Considering the definitions given above, it is certainly unnatural from many points of view that the input has to be fed to a specific level of the tree. From the VLSI point of view, this means that there is also a lower bound on the size of words that can be processed by a chip—an upper bound is obvious. Hence, it is desirable that an arbitrary STA can always be replaced by an equivalent one that is **stable** in the sense that the accepted language remains invariant even if we drop the condition that the input word must always be fed on the first possible level of the tree. The "first possible level" was expressed in our previous definitions by saying that k is the *smallest* integer such that LEVEL(k) in T contains at least as many nodes as the length of the input word indicates.

Thus, in a stable STA an input word can be fed on any sufficiently long level; the acceptance of the word is independent of the level. The following result, the proof of which is omitted, is pleasing because it shows that our previous somewhat awkward input condition is not actually needed.

Theorem 8.13.   For every systolic tree automaton, an equivalent stable systolic tree automaton can be constructed.

It is fairly easy to decide the emptiness of the language accepted by a systolic tree automaton. By the second sentence of Theorem 8.12, this also leads to the decidability of the equivalence problem for systolic tree automata. As a corollary of this result, it is possible to obtain the decidability of stability.

The classical theory of tree automata lies beyond the scope of this book; the reader is referred to [GeS]. There are similarities, as well as also definite differences, between classical and systolic tree automata.

So far, no explicit characterization is known for the family $\mathcal{L}$ of languages accepted by systolic tree automata. We know that $\mathcal{L}$ contains every regular language and, in addition, non-context-free languages where the word lengths grow exponentially. (See Example 8.5 and Theorem 8.12.) In addition, $\mathcal{L}$ is closed under Boolean operations.

It is more difficult to give examples of languages in $\mathcal{L}$ not obtainable by these rules—for instance, examples of context-free nonregular languages in $\mathcal{L}$. We now give such an example, due to [Pat].

Example 8.6. Let $L \subseteq \{a, b\}^*$ be the language consisting of all words that are not of the form

$$bbabaaab\cdots a^{2^r-1}b, \qquad r \geq 0. \tag{4}$$

Hence, the complement of L, $\sim L$, consists of all words of the form (4).

We prove first that L is acceptable by a systolic tree automaton. By Theorem 8.12, it suffices to construct a systolic tree automaton STA accepting $\sim L$. STA will be in the normal form of Theorem 8.11 and possesses the operating alphabet $\{a, b, \#, C, D\}$, where D is the only accepting letter. The functions g and f are defined by

$$g(x) = x \qquad \text{for } x = a, b, \#;$$
$$f(a, a) = a, \qquad f(a, b) = C, \qquad f(a, C) = C,$$
$$f(b, b) = f(D, C) = D,$$
$$f(x, y) = \# \qquad \text{in other cases.}$$

Examples of acceptance and nonacceptance are given in Figure 8.16. Since g is the identity, the lowest line gives the input directly.

The reader should have no difficulties in verifying that this STA accepts $\sim L$. Observe, in particular, that the length of every word in $\sim L$ is a power of 2. On the other hand, the presence of # anywhere causes the rejection of the word. Hence, the automaton STA is not stable: Every word of $\sim L$ can be accepted only from the correct level. The reader might want to construct an equivalent stable automaton.

It is obvious that neither L nor $\sim L$ is regular. On the other hand, L is context-free. We leave the construction of a context-free grammar generating L to the reader. Clearly, $\sim L$ is not context-free.   □

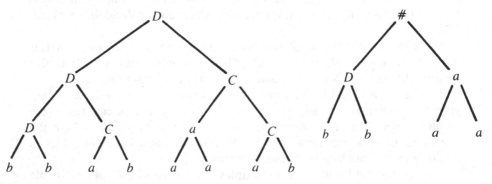

FIGURE 8.16

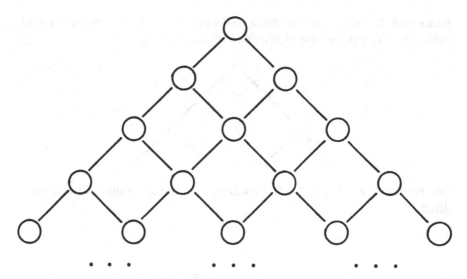

FIGURE 8.17

For a **systolic trellis automaton,** the underlying structure is an infinite trellis, as in Figure 8.17.

All formal details are similar for tree and trellis automata. Therefore, we give only a brief intuitive description of trellis automata. (If the diagonal connections are added, then a trellis can obviously be viewed as a hexagonal grid.)

The nodes of the underlying trellis are labeled. If only one label is used (which means, intuitively, that the processors are all identical), the trellis is referred to as **homogeneous.** The trellis is **semihomogeneous** iff it has only finitely many infinite labeled subtrellises. The trellis is **regular** iff the labeling is top-down deterministic: The label of a daughter node is uniquely determined by the label(s) of its mother(s).

As in connection with tree automata, two functions g_A and f_A are associated with each label A. All data paths are bottom-up. The input is fed at the level possessing the length of the input. (Observe that there are levels of all lengths. We do not consider the empty word as an input. Hence, the filling marker # is not needed.) The value of f_A is sent upward to one or two nodes. More specifically, if the node labeled by A lies on the left or right "leg" of the trellis, it has only one mother that receives the value of f_A. Otherwise, it has two mothers, both receiving this value. As before, the acceptance depends on the output from the root of trellis.

Example 8.7.   First, observe that a trellis can be semihomogeneous without being regular. An example is provided by the labeling

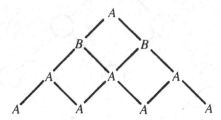

On the other hand, the trellis *T* below is regular, but all its subtrellises are still distinct:

$$
\begin{array}{ccccccccc}
& & & & A & & & & \\
& & & A & & A & & & \\
& & A & & B & & A & & \\
& A & & A & & A & & A & \\
A & & B & & B & & B & & A \\
\end{array}
$$

A   A   B   B   A   A

A   B   A   B   A   B   A

A A A A A A A A

.

.

.

The deterministic top-down rule is here very simple: A daughter is labeled by *A* if it has only one mother or two distinctly labeled mothers; it is labeled by *B* if it has two identically labeled mothers.

A trellis automaton for the language

$$L = \{a^{2^n} \mid n \geq 0\},$$

with *T* as its underlying trellis, is fairly easy to construct. On the other hand, it can be shown that *L* is not accepted by any semihomogeneous trellis automaton. Every language accepted by a semihomogeneous trellis automaton is accepted also by a homogeneous one. (This result corresponds to Theorem 8.11.) The family of languages accepted by regular trellis automata is strictly larger than that accepted by homogeneous ones.

It can also be shown that the family of languages accepted by systolic tree automata is contained in the family of languages accepted by homogeneous trellis automata. That the containment is strict is seen by considering the Dyck language D_2 of properly nested words from two pairs of parentheses: (,) and [,]. More specifically, D_2 is the language over the alphabet {(,), [,]} generated by the context-free grammar

$$S \rightarrow SS, \qquad S \rightarrow (S), \qquad S \rightarrow [S], \qquad S \rightarrow (), \qquad S \rightarrow [].$$

It is fairly easy to prove that D_2 is not accepted by any systolic tree automaton. However, the following semihomogeneous trellis automaton A accepts D_2. By the result mentioned above, A can be replaced by a homogeneous trellis automaton.

The nodes of the underlying trellis are labeled by r (root), i (all interior nodes), a, and b (nodes on the left and right leg, except the root). The operating alphabet of A is $\{(,), [,], Y, N\}$, where Y is the only accepting letter. Each of the input functions g is the identity function. The functions f are defined by the following tables. The first variable of f is read from the rows and the second from the columns.

f_r	(	)	[	]	Y	N
(	N	Y	N	N	N	N
)	N	N	N	N	N	N
[	N	N	N	Y	N	N
]	N	N	N	N	N	N
Y	N	N	N	N	Y	N
N	N	N	N	N	N	N

f_i	(	)	[	]	Y	N
(	(	Y	(	N	(	N
)	Y	)	Y	]	Y	N
[	[	N	[	Y	[	N
]	Y	)	Y	]	Y	N
Y	Y	)	Y	]	Y	N
N	N	N	N	N	N	N

f_a	(	)	[	]	Y	N
(	(	Y	(	N	(	N
)	N	N	N	N	N	N
[	[	N	[	Y	[	N
]	N	N	N	N	N	N
Y	Y	)	Y	]	Y	N
N	N	N	N	N	N	N

f_b	(	)	[	]	Y	N
(	N	Y	N	N	N	N
)	N	)	N	]	Y	N
[	N	N	N	Y	N	N
]	N	)	N	]	Y	N
Y	N	)	N	]	Y	N
N	N	N	N	N	N	N

Intuitively, the automaton A transmits left parentheses (and [(resp. right parentheses) and]) along left (resp. right) obliques until they either match properly, hit the left or right leg without being matched, or else match improperly. In the first case, the output Y is produced. The output N is produced in the other cases. Outputs Y are properly absorbed, but outputs N can never be eliminated.

The following trellis is a typical example of acceptance.

```
                Y
              (   )
            (   Y   )
          (   (   )   )
        (   (   ]   )   )
      (   (   Y   Y   )   )
    (   (   Y   Y   Y   )   )
  (   (   (   )   [   ]   )   )
```
□

Some results about trellis automata, notably results about the comparison of different types of trellis automata and tree automata, were already mentioned in Example 8.7. We list, finally, some further results.

For every homogeneous trellis automaton, an equivalent stable one can be constructed. In the stable automaton, an input w can be fed at any level whose length is greater than or equal to $|w|$. If necessary, we use the marker #; the input becomes $w\#^i$.

Moreover, for every homogeneous trellis automaton, there is an equivalent *superstable* one. This means that an input w can be fed at any sufficiently long level in such a way that the markers # are positioned arbitrarily—that is, not necessarily at the end of w. For instance, the word *aba* can be fed at the level of length seven in the form $a\#b\#\#a\#$ or in the form $\#a\#^3ba$. The corresponding superstability result does not hold for tree automata because one can show that every language accepted by a superstable systolic tree automaton is regular.

Emptiness and equivalence problems are undecidable for homogeneous trellis automata. It is also undecidable whether or not a given homogeneous trellis automaton is stable (resp. superstable).

The family of languages accepted by homogeneous trellis automata contains all linear languages and is closed under Boolean operations.

In a natural way, one can also define **nondeterministic** variants of systolic tree and trellis automata. It turns out that nondeterministic tree automata can be simulated by deterministic ones, but the same does not hold true in regard to trellis automata. Some naturally restricted Turing machine models can be constructed for various types of deterministic and nondeterministic trellis automata. For instance, the family of languages accepted by nondeterministic homogeneous trellis automata equals the family of languages accepted by certain restricted Turing machines. Such Turing machine characterizations also yield a number of complexity results.

EXERCISES

8.1. Are the languages $L(PN_2, M)$ and $L(PN_3, M)$ context-free for all M? (The Petri nets PN_2 and PN_3 are the ones defined above.) If they are, modify the Petri nets in such a way that non-context-free languages result.

8.2. Give a characterization for the family of languages of the form $L(PN, M)$. (See [St]. We mean here a characterization of the following type: The family consists of all languages obtainable from certain finitely many "start" languages by certain operations.)

8.3. Prove Theorem 8.3.

8.4. Prove the last sentence in Section 8.1.

8.5. Prove Theorem 8.6.

8.6.   Show that the grammar determined by the productions

$$S \to aSS, \qquad S \to aS, \qquad S \to a$$

   is complete. (A very elegant recent proof of this result is contained in
   [EhR]. The resulting normal form for context-free grammars is
   known as the **Greibach normal form**.)

8.7.   Prove that every grammatical family is closed under intersection with
   regular languages. In other words, if L is a language in the family and
   R is a regular language, then $L \cap R$ is in the family.

8.8.   Prove Theorems 8.11 and 8.13.

8.9.   Show that the equivalence problem for systolic tree automata is decidable.

8.10.   Show that D_2 is not acceptable by a systolic tree automaton.

8.11.   Show that emptiness and equivalence problems are undecidable for
   homogeneous trellis automata. Show also that every linear language
   is acceptable by a homogeneous trellis automaton. ([CuGS3] and
   [CuGS4] can be consulted.)

8.12.   Modify STA in such a way that an arbitrary underlying tree is
   allowed, provided only finitely many letters are used in the labeling
   of the nodes. Prove that every language whatsoever over the one-
   letter alphabet $\{a\}$ is acceptable by such a device.

8.13.   Our remaining exercises deal with some currently very active notions
   and topics not discussed in Chapter 8. The definitions will only be
   outlined. We begin with a general model for grammatical considerations, called a **selective substitution grammar**, SSG. This model was
   introduced in [Roz1]. Detailed information, as well as a comprehensive list of references, is contained in [Kle]. The reader is referred also
   to [Roz2], where a more-general model, covering both automata and
   grammars, is introduced.

   Essentially, an SSG consists of a context-free grammar G and
   of a language K over the alphabet $\Sigma \cup \bar{\Sigma}$, where Σ is the total alphabet
   (nonterminals and terminals) of G and $\bar{\Sigma}$ consists of "barred" versions of letters in Σ. The language K is referred to as the **selector** for
   rewriting. The barred letters in words of K indicate the (occurrences
   of the) symbols to be rewritten during one rewriting step. For instance, if the word $a\bar{A}BA\bar{B}$ in K is used when rewriting the word
   $aABAB$, then the first A and the second B must be rewritten. Observe, however, that several words of the selector might be applicable. The selector might contain, for instance, also the word $a\bar{A}BAB$,
   which gives rise to another possibility of rewriting in this case.

   Thus, a selector of the form $\Sigma^* \bar{\Sigma} \Sigma^*$ gives rise to an ordinary
   context-free grammar. A selector of the form $\bar{\Sigma}^*$ gives rise to an $0L$ or
   $E0L$ system.

   Construct an SSG, as simple as possible, generating the language $\{a^n b^n c^n \mid n \geq 1\}$.

Prove that any language whatsoever is generated by an SSG. (Here the complexity of the language is reflected in the complexity of the selector.)

Study restrictions of SSG's such that the resulting language family is the family of context-free languages. (The restriction of the selector being of the form $\Sigma^*\bar{\Sigma}\Sigma^*$ is not the only possible one.)

8.14.   An important special case of an SSG is a **pattern grammar**: The rewriting of a symbol depends on its position only. More specifically, the selector is a language over the alphabet $\{0, 1\}$. The letter 1 indicates the symbols to be rewritten. For instance, if 00110 is in the selector, then in any word of length 5, the third and fourth letter may be rewritten in one derivation step. On the other hand, if this word of the selector is used, then both the third and the fourth letter *must* be rewritten in one derivation step.

Regular pattern grammars refer to the case where the selector is a regular language over the alphabet $\{0, 1\}$.

Give examples of non-context-free languages generated by regular pattern grammars.

Show that the family of *E0L* languages is properly contained in the family of languages generated by regular pattern grammars.

Show also that every recursively enumerable language L is a morphic image of a language generated by a regular pattern grammar.

Strengthen the latter result to the following form: If # is not in the alphabet of L, then $L\#^5$ is generated by a regular pattern grammar.

8.15.   The words and languages discussed in this book are essentially one-dimensional. Many-dimensional aspects of the theory have been investigated to a large extent. In particular, the theory of **graph grammars** constitutes today a highly developed body of knowledge. Applications of graph grammars range from syntax and semantics of programming languages, compiler techniques, data bases, data-flow analysis, and concurrency to pattern recognition and developmental biology.

Consider node-labeled, undirected graphs. A production associates to a label A a graph G. One rewriting step consists of replacing a node labeled with A by the graph G. Moreover, a **connection relation** over $\Sigma \times \Sigma$ is given, indicating which of the nodes of G are connected with the earlier neighbors of the A-node. For instance, if a node labeled by C is originally a neighbor of our A-node and G contains a node labeled by B and, moreover, the pair (B, C) belongs to the connection relation, then the C-node and the B-node are connected after the rewriting step.

To repeat: One rewriting step consists of replacing an A-node by a graph G, in accordance with some production. To complete the rewriting step, some nodes of G are connected with the original neighbors of the A-node in accordance with the connection relation. The formal details can be found in [JaR2]. The references [JaR1] and [EhrNR] contain more information about graph grammars.

Construct a graph grammar generating all node-labeled graphs. Construct a graph grammar generating all 3-colorable graphs.

8.16. Give an example of a family of graphs that cannot be generated by any graph grammar.

8.17. Show that the equivalence of two graph grammars is undecidable (that is, the problem of whether or not the families of generated graphs coincide).

8.18. Study representations of context-free languages in terms of graph grammars.

Historical and
Bibliographical Remarks

The subsequent remarks, as well as the references in this book in general, are intended to aid the reader rather than to credit each individual result to some specific author(s). Indeed, because of the diversity of the presented material, such a detailed tracing of the origins of individual results would be rather frustrating. On the other hand, the references try to give proper credit to the major developments and results. They also try to give hints on some specific details to the interested reader.

The papers by Thue, of which [Th1] and [Th2] are perhaps the most important ones, can be viewed as the first contribution to language theory. Rewriting systems are also often referred to as *semi-Thue systems*. (Here *semi* refers to the rewriting rules being nonsymmetric: Thue considered systems where $u \to w$ and $w \to u$ are always together.) The notion of a grammar was introduced in [Ch1] and [Ch2], whereas [ChS] and [Ba-H] are early papers in the theory of context-free languages. [Gi1] contains the details and references. Indeed, there is a huge literature concerning matters related to Chapter 2. [Ber], [Har], [HoU], and [Sa3] and their references may be consulted. Example 2.2 is essentially due to [MacP] and Example 2.5 to M. Soittola ([Sa3]). Post systems were introduced in [Po1], whereas regular systems are due to [Bu]. Our exposition largely follows [Sa2]. [Mar] is the best source for Markov algorithms. L systems were introduced in [Li]; [RozS] contains a detailed exposition.

Kleene characterization was first given in the basic paper [Kl]. [Ei], [MacP], and [Sa2] contain details of various aspects of finite automata theory. [Paz] discusses the corresponding probabilistic aspects. [Gi2] may be consulted for more general aspects of gsm theory, as well as for the AFL-theory, which investigate closure properties of language families and are beyond the scope of this book. [MSW5], [CuS], and [Ho] are the basic papers about number systems. The pumping lemma of context-free languages is due to [Ba-H], and Theorem 3.14 to [Ja]. [Gi1], [Har], [HoU], and [Ku] may be consulted in regard to pushdown automata.

[Ro] constitutes an excellent reference concerning the matters discussed in Chapter 4. For many details about Turing machines, [Da1] and [HoU] may also be consulted. Of course, the basic ideas about undecidability and indexing go back to the seminal paper [Go]. Turing machines were introduced in [Tu]. Much of the history of recursive functions is traced in [Ro]. Rice's theorem was established in [Ri]. Examples 4.2 and 4.4 are essentially due to J. Honkala and Sheng Yu, and Section 4.6 is modified from [Sa1].

The Post correspondence problem was introduced in [Po2]. The reader is referred to [Cl], [EhKR], [Ru3], and [Sa5] for recently modified aspects of the problem. In regard to Theorem 5.6, the corresponding equivalence problem for D0L systems was open for a long time, until the decidability was established in [CuF]. Hilbert's tenth problem was solved in [Mat] after many developments due to a number of people. Details can be found in [Da2]; [Ru2] may also be consulted. The decidability of the reachability problem for vector addition systems was established in [May]; a simplified version is due to [Kos].

The books [AHU], [GaJ], [HoU], and [MaY] present various aspects of complexity theory. [HarS] is a basic paper in the theory, and [Bl1] introduces the axiomatic aspect and also contains (maybe in a somewhat different form) many of the theorems of Sections 6.1–6.3. The gap theorem (Theorem 6.8) is due to [Bo]. Theorem 6.12 is due to [Da3]. [So] and [BlM] contain more information about speedability, and [Be] has details of the matters discussed at the end of Section 6.3. $\mathcal{NP}$-completeness was introduced in [Co], [Kar] being a basic paper concerning different types of $\mathcal{NP}$-complete problems. For sparse sets, the reader is referred to [HarB] and [Mah]. [MeS] is a fundamental paper in the area of provably intractable problems.

[Kah] and [Ga] can be used as an introduction to classical cryptosystems. Fundamentals of cryptography from the point of view of information theory were developed in [Shan]. [Mau] gives a good overview of some practical aspects. Public key systems were initiated in [DH], and the books [De] and [Ko] contain additional material. [Bet] is a many-sided survey. Example 7.3 is due to [Hil], and the essential part of Example 7.5 to [Ham]. The importance of commutativity in cryptography is discussed in [Sh2]. Example 7.7 is due to [SaW], and Theorem 7.1 to [Br]. Knapsack systems were introduced in [MeH], and the signature aspect was discussed in [Sh1]. [EY] is an interesting contribution to the complexity aspect. Examples 7.9 and 7.9a are based on the work of Shamir: the former on the famous paper [Sh3], and the latter on [Sh4]. RSA system was introduced in [RSA]. Example 7.11 is based on [Ra2]. [Ra1] is basic in probabilistic primality algorithms. In regard to mental poker (Example 7.12), the reader is referred to [ShRA] and [Lip], the latter concerning cheating with quadratic residues. [B12] and [Y] may be consulted in connection with Examples 7.13 and 7.14, respectively.

Petri nets were introduced by Petri; for instance, see [Pe]. By now, there are many books dealing with the topic. [RozV] discusses subset languages mentioned at the end of Section 8.1. [Wo] constitutes a comprehensive exposition on grammar forms, [CrG] being the original paper in this area. The material in Section 8.2 is drawn from [MSW1]–[MSW4], [MSW6] and [We]. The basics of systolic tree automata are developed in [CuSW], [Ste], [CuGS1], and [CuGS2], and those of systolic trellis automata in [CuGS3] and [CuGS4]. Complexity results are established in [IK], Example 8.6 being due to [Pat].

The following list of references is not intended to be a comprehensive bibliography. It contains only the books and papers to which this book referred. Many of the books listed contain large bibliographies.

References

[AHU]   A. Aho, J. Hopcroft, and J. Ullman. *The Design and Analysis of Computer Algorithms.* Reading, Mass.: Addison-Wesley (1974).

[Ba-H]   Y. Bar-Hillel. *Language and Information.* Reading, Mass.: Addison-Wesley (1964).

[Be]   V. L. Bennison. Recursively enumerable complexity sequences and measure independence. *Journal of Symbolic Logic* 45(1980): 417–438.

[Be-OCS]   M. Ben-Or, B. Chor, and A. Shamir. On the cryptographic security of single RSA bits. *Proceedings of the 15th ACM Symposium on the Theory of Computing* (1983): 421–430.

[Ber]   J. Berstel. *Transductions and Context-Free Languages.* Stuttgart: B. G. Teubner (1979).

[Bet]   T. Beth (ed.). Cryptography. *Springer Lecture Notes in Computer Science* 149(1983).

[Bl1]   M. Blum. A machine-independent theory of the complexity of recursive functions. *Journal of the Association for Computing Machinery* 14(1967): 322–336.

[Bl2]   M. Blum. Coin flipping by telephone. A protocol for solving impossible problems. *SIGACT News* (1981): 23–27.

[Bl3] M. Blum. How to exchange (secret) keys. *Proceedings of the 15th ACM Symposium on the Theory of Computing* (1983): 440–447.

[BlM] M. Blum and I. Marques. On complexity properties of recursively enumerable sets. *Journal of Symbolic Logic* **38**(1973): 579–593.

[Bo] A. Borodin. Computational complexity and the existence of complexity gaps. *Journal of the Association for Computing Machinery* **19**(1972): 158–174.

[Br] G. Brassard. Relativized cryptography. *Proceedings of the 20th Symposium FOCS* (1979): 383–391.

[Bu] J. R. Büchi. Regular canonical systems. *Archiv für Mathematische Logik und Grundlagenforschung* **6**(1964): 91–111.

[Ch1] N. Chomsky. Three models for the description of language. *IRE Transactions on Information Theory* **IT-2**(1956): 113–124.

[Ch2] N. Chomsky. *Syntactic Structures.* Gravenhage: Mouton (1957).

[ChS] N. Chomsky and M. P. Schützenberger. The algebraic theory of context-free languages. In P. Braffort and D. Hirschberg (eds.), *Computer Programming and Formal Systems,* Amsterdam: North-Holland Publ. (1963): 118–161.

[Cl] V. Claus. Die Grenze zwischen Entscheidbarkeit und Nichtentscheidbarkeit. *Fernstudienkurs für die Fernuniversität Hagen.* Hagen: Open University of Hagen (1979).

[Co] S. A. Cook. The complexity of theorem-proving procedures. *Proceedings of the 3rd Annual ACM Symposium on the Theory of Computing,* New York: Association for Computing Machinery (1971): 151–158.

[CrG] A. Cremers and S. Ginsburg. Context-free grammar forms. *Journal of Computer and System Sciences* **11**(1975): 86–116.

[CuF] K. Culik II and I. Fris. The decidability of the equivalence problem for D0L-systems. *Information and Control* **35**(1977): 20–39.

[CuGS1] K. Culik II, J. Gruska, and A. Salomaa. Systolic automata for VLSI on balanced trees. *Acta Informatica* **18**(1983): 335–344.

[CuGS2] K. Culik II, J. Gruska, and A. Salomaa. On a family of *L* languages resulting from systolic tree automata. *Theoretical Computer Science* **23**(1983): 231–242.

[CuGS3] K. Culik II, J. Gruska, and A. Salomaa. Systolic trellis automata. *International Journal of Computer Mathematics* **15**(1984): 195–212 and **16**(1984): 3–22.

[CuGS4]   K. Culik II, J. Gruska, and A. Salomaa. Systolic trellis automata: stability, decidability and complexity. *Information and Control,* to appear.

[CuS]   K. Culik II and A. Salomaa. Ambiguity and decision problems concerning number systems. *Information and Control* 56(1984): 139-153.

[CuSW]   K. Culik II, A. Salomaa, and D. Wood. Systolic tree acceptors. *RAIRO* 18(1984): 53-69.

[Da1]   M. Davis. *Computability and Unsolvability.* New York: McGraw-Hill (1958).

[Da2]   M. Davis. Hilbert's tenth problem is unsolvable. *American Mathematical Monthly* 80(1973): 233-269.

[Da3]   M. Davis. Speed-up theorems and Diophantine equations. In R. Rustin (ed.), *Computational Complexity,* New York: Algorithmics Press (1973): 87-95.

[De]   D. E. Denning. *Cryptography and Data Security.* Reading, Mass.: Addison-Wesley (1982).

[DH]   W. Diffie, and M. Hellman. New directions in cryptography. *IEEE Transactions on Information Theory* IT-22(1976): 644-654.

[EhKR]   A. Ehrenfeucht, J. Karhumäki, and G. Rozenberg. The (generalized) Post correspondence problem with lists consisting of two words is decidable. *Theoretical Computer Science* 21(1982): 119-144.

[EhR]   A. Ehrenfeucht and G. Rozenberg. On Greibach normal form. Manuscript (1983).

[EhrNR]   H. Ehrig, M. Nagl, and G. Rozenberg (ed.). Graph-Grammars and Their Application to Computer Science. *Springer Lecture Notes in Computer Science* 153(1983).

[Ei]   S. Eilenberg. *Automata, Languages and Machines,* Vol. A. New York: Academic Press (1974).

[EY]   S. Even and Y. Yacobi. Cryptosystems which are NP-hard to break. Computer Science Department, Haifa: Technion (1979).

[Ga]   H. F. Gaines. *Cryptanalysis.* New York: Dover Publications (1939).

[GaJ]   M. R. Garey and D. S. Johnson. *Computers and Intractability,* San Francisco: W. H. Freeman (1979).

[GeS]   F. Gécseg and M. Steinby. *Tree Automata.* Budapest: Akadémiai Kiadó (1982).

[Gi1]   S. Ginsburg. *The Mathematical Theory of Context-Free Languages.* New York: McGraw-Hill (1966).

[Gi2]   S. Ginsburg. *Algebraic and Automata-Theoretic Properties of Formal Languages.* Amsterdam: North-Holland Publ. (1975).

[Go]   K. Gödel. Über formal unentscheidbare Sätze der Principia Mathematica und verwandter Systeme I. *Monatshefte für Mathematik und Physik* **38**(1931): 173–198.

[GM]   S. Goldwasser and S. Micali. Probabilistic encryption and how to play mental poker keeping secret all partial information. *Proceedings of the 14th ACM Symposium on the Theory of Computing* (1982): 365–377.

[Ham]   C. Hammer. High order homophonic ciphers. *Cryptologia* 5(1981): 231–242.

[Har]   M. Harrison. *Introduction to Formal Language Theory.* Reading, Mass.: Addison-Wesley (1978).

[HarB] J. Hartmanis and L. Berman. On polynomial time isomorphisms of some new complete sets. *Journal of Computer and System Sciences* **16**(1978): 418–422.

[HarS] J. Hartmanis and R. E. Stearns. On the computational complexity of algorithms. *Transactions of the American Mathematical Society* **117**(1965): 285–306.

[Hi]   D. Hilbert. Mathematische Probleme. Vortrag, gehalten auf dem internationalen Mathematiker-Kongress zu Paris 1900. *Gesammelte Abhandlungen* III. Springer, Berlin (1935).

[Hil]   L. S. Hill. Cryptography in an algebraic alphabet. *American Mathematical Monthly* **36**(1929): 306–312.

[Ho]   J. Honkala. Unique representation in number systems and L codes. *Discrete Applied Mathematics* 4(1982): 229–232.

[HoP]   J. Hopcroft and J. J. Pansiot. On the reachability problem for 5-dimensional vector addition systems. *Theoretical Computer Science* **8**(1979): 135–159.

[HoU]   J. Hopcroft and J. Ullman. *Introduction to Automata Theory, Languages and Computation.* Reading, Mass.: Addison-Wesley (1979).

[IK]   O. H. Ibarra and S. M. Kim. Characterizations and computational complexity of systolic trellis automata. *Theoretical Computer Science* **29**(1984): 123–154.

[Ja]   J. Jaffe. A necessary and sufficient pumping lemma for regular languages. *SIGACT News* (1978): 48–49.

[JaR1] D. Janssens and G. Rozenberg. On the structure of node-labeled controlled graph languages. *Information Sciences* **20**(1980): 191–216.

[JaR2] D. Janssens and G. Rozenberg. Graph grammars with node-labeled controlled rewriting and embedding. *Springer Lecture Notes in Computer Science* **153**(1983): 186–205.

[Jo]   J. P. Jones. Universal Diophantine equation. *Journal of Symbolic Logic* **47**(1982): 549–571.

[JoSWW] J. P. Jones, D. Sato, H. Wada, and D. Wiens. Diophantine presentation of the set of prime numbers. *American Mathematical Monthly* **83**(1976): 449–464.

[Kah]   D. Kahn. *The Codebreakers: The Story of Secret Writing*, New York: Macmillan (1967).

[Kar]   R. M. Karp. Reducibility among combinatorial problems. In R. E. Miller and J. W. Thatcher (eds.), *Complexity of Computer Computations*, New York: Plenum Press (1972): 85–103.

[KeH]   C. F. Kent and B. R. Hodgson. An arithmetical characterization of NP. *Theoretical Computer Science* **21**(1982): 255–267.

[Kh]   L. G. Khachiyan. A polynomial algorithm for linear programming. *Doklady Akademii Nauk SSSR* **244**(1979): 1093–1096.

[Kl]   S. C. Kleene. Representation of events in nerve nets and finite automata. In: *Automata Studes*, Princeton: Princeton University Press (1956): 3–42.

[Kle]   H. C. M. Kleijn. Selective substitution grammars based on context-free productions. Dissertation, Leiden University (1983).

[Ko]   A. Konnheim. *Cryptography: A Primer*. New York: Wiley and Sons (1982).

[Kos]   S. R. Kosaraju. Decidability of reachability in vector addition systems. *Proceedings of the 14th Annual ACM Symposium on the Theory of Computing* (1982): 267–281.

[Ku]   W. Kuich. Formal power series, cycle-free automata and algebraic systems. Institut für Informationsverarbeitung, TU Graz, Bericht F **103** (1982).

[LeP]   H. R. Lewis and C. H. Papadimitriou. *Elements of the Theory of Computation*. Englewood Cliffs, N.J.: Prentice-Hall (1981).

[Li]   A. Lindenmayer. Mathematical models for cellular interaction in development I-II. *Journal of Theoretical Biology* **18**(1968): 280-315.

[Lip]   R. J. Lipton. How to cheat at mental poker. Berkeley, Calif.: Computer Science Department, University of California (1979).

[MaY]   M. Machtey and P. Young. *An Introduction to the General Theory of Algorithms*. New York: Elsevier North-Holland (1978).

[MacP]   R.McNaughton and S. Papert. *Counter-Free Automata*. Cambridge, Mass.: M.I.T. Press (1971).

[Mah]   S. R. Mahaney. Sparse complete sets for NP: solution of a conjecture of Berman and Hartmanis. *Journal of Computer and System Sciences* **25**(1982): 130-143.

[Mar]   A. A. Markov. *Theory of Algorithms*. Jerusalem: Israel Program for Scientific Translations (1961).

[Mat]   J. Matijasevich. Diofantovo predstavlenie perechislimykh predikatov. *Izvestija Akademii Nauk SSSR, Ser. Matem.* **35**(1971): 3-30.

[Mau]   H. A. Maurer. On some developments in cryptography and their applications to computer science. Institut für Informationsverarbeitung, TU Graz, Bericht 31 (1979).

[MSW1]   H. A. Maurer, A. Salomaa, and D. Wood. Context-free grammar forms with strict interpretations. *Journal of Computer and System Sciences* **21**(1980): 110-135.

[MSW2]   H. A. Maurer, A. Salomaa, and D. Wood. Finitary and infinitary interpretations of languages. *Mathematical Systems Theory* **15**(1982): 251-265.

[MSW3]   H. A. Maurer, A. Salomaa, and D. Wood. Dense hierarchies of grammatical families. *Journal of the Association for Computing Machinery* **29**(1982): 118-126.

[MSW4]   H. A. Maurer, A. Salomaa, and D. Wood. Colorings and interpretations: a connection between graphs and grammar forms. *Discrete Applied Mathematics* **3**(1981): 119-135.

[MSW5]   H. A. Maurer, A. Salomaa, and D. Wood. L codes and number systems. *Theoretical Computer Science* **22**(1983): 331-346.

[MSW6]   H. A. Maurer, A. Salomaa, and D. Wood. A supernormal-form theorem for context-free grammars. *Journal of the Association for Computing Machinery* **30**(1983): 95-102.

[May]   E. W. Mayr. An algorithm for the general Petri net reachability problem. *Proceedings of the 13th Annual ACM Symposium on the Theory of Computing* (1981): 238-246.

[MeH]   R. Merkle and M. Hellman. Hiding information and signatures in trapdoor knapsacks. *IEEE Transactions on Information Theory* **IT-24**(1978): 525–530.

[MeS]   A. R. Meyer and L. J. Stockmeyer. The equivalence problem for regular expressions with squaring requires exponential time. *Proceedings of the 13th Annual Symposium on Switching and Automata Theory*, Long Beach, Calif.: IEEE Computer Society (1972): 125–129.

[Pa]   J. J. Pansiot. A note on Post's correspondence problem. *Information Processing Letters* **12**(1981): 233–234.

[Pat]   M. Paterson. Solution to P8, Number 17. *EATCS Bulletin* **18** (1982): 29.

[Paz]   A. Paz. *Introduction to Probabilistic Automata.* New York: Academic Press (1971).

[Pe]   C. A. Petri. Kommunikation mit Automaten. *Schriften des. Inst. für Instrumentelle Mathematik* Nr. 2, Bonn (1962).

[Pi]   N. Pippinger. Pebbling. *Proceedings of the 5th IBM Symposium on Mathematical Foundations of Computer Science,* IBM Japan (1980): 1–19.

[Po1]   E. L. Post. Formal reductions of the general combinatorial decision problem. *American Journal of Mathematics* **65**(1943): 197–215.

[Po2]   E. L. Post. A variant of a recursively unsolvable problem. *Bulletin of the American Mathematical Society* **52**(1946): 264–268.

[Ra1]   M. O. Rabin. Probabilistic algorithms. In, J. Traub (ed.) *Algorithms and Complexity, New Directions and Recent Results,* New York: Academic Press (1976): 21–40.

[Ra2]   M. O. Rabin. Digitalized signatures and public-key functions as intractable as factorization. Cambridge, Mass.: MIT Laboratory for Computer Science (1979).

[Ri]   H. G. Rice. Classes of recursively enumerable sets and their decision problems. *Transactions of the American Mathematical Society* **74**(1953): 358–366.

[RSA]   R. Rivest, A. Shamir and L. Adleman. A method for obtaining digital signatures and public-key cryptosystems. *ACM Communications* **21**(1978): 120–126.

[Ro]   H. Rogers, Jr. *Theory of Recursive Functions and Effective Computability.* New York: McGraw-Hill (1967).

[Roz1] G. Rozenberg. Selective substitution grammars. *Elektronische Informationsverarbeitung und Kybernetik* **13**(1977): 455–463.

[Roz2] G. Rozenberg. On coordinated selective substitutions. Manuscript (1983).

[RozS] G. Rozenberg and A. Salomaa. *The Mathematical Theory of L Systems.* New York: Academic Press (1980).

[RozV] G. Rozenberg and R. Verraedt. Subset languages of Petri nets. *Proceedings of the 3rd European Workshop on Applications and Theory of Petri Nets* (1982): 407–420.

[Ru1]   K. Ruohonen. Hilbertin kymmenes probleema. *Arkhimedes* **24**(1972): 71–100.

[Ru2]   K. Ruohonen. Hilberts tionde problem. *Nordisk Matematisk Tidskrift* **28**(1980): 145–154.

[Ru3]   K. Ruohonen. On some variants of Post's correspondence problem. *Acta Informatica* **19**(1983): 357–367.

[Sa1]   A. Salomaa. Some analogues of Sheffer functions in infinite-valued logics. *Acta Philosophica Fennica* **16**(1963): 227–235.

[Sa2]   A. Salomaa. *Theory of Automata.* Oxford: Pergamon Press (1969).

[Sa3]   A. Salomaa. *Formal Languages.* New York: Academic Press (1973).

[Sa4]   A. Salomaa. On sentential forms of context-free grammars. *Acta Informatica* **2**(1973): 40–49.

[Sa5]   A. Salomaa. *Jewels of Formal Language Theory.* Rockville: Computer Science Press (1981).

[SaS]   A. Salomaa and M. Soittola. *Automata-Theoretic Aspects of Formal Power Series.* Berlin: Springer-Verlag (1978).

[SaW]   A. Salomaa and E. Welzl. A cryptographic trapdoor based on iterated morphisms. Manuscript (1983).

[Sav]   W. Savitch. How to make arbitrary grammars look like context-free grammars. *SIAM Journal on Computing* **2**(1973): 174–182.

[Sh1]   A. Shamir. A fast signature scheme. Cambridge, Mass.: MIT Laboratory for Computer Science (1978).

[Sh2]   A. Shamir. On the power of commutativity in cryptography. *Springer Lecture Notes in Computer Science* **85**(1980): 582–595.

[Sh3]   A. Shamir. A polynomial time algorithm for breaking the basic Merkle-Hellman cryptosystem. *Proceedings of the 23rd FOCS Symposium* (1982): 145–152.

[Sh4]   A. Shamir. Embedding cryptographic trapdoors in arbitrary knapsack systems. Manuscript (1983).

[ShRA]   A. Shamir, R. Rivest and L. Adleman. Mental poker. In, D. A. Klarner (ed.), *The Mathematical Gardner*. Belmont, Calif.: Wadsworth International (1981): 37–43.

[Shan]   C. E. Shannon. Communication theory of secrecy systems. *Bell System Technical Journal* **28**(1949): 656–715.

[So]   R. I. Soare. Computational complexity, speedable and levelable sets. *Journal of Symbolic Logic* **42**(1977): 545–563.

[St]   P. H. Starke. *Petri-Netze*. Berlin: VEB Deutscher Verlag der Wissenschaften (1980).

[Ste]   M. Steinby. Systolic trees and systolic language recognition by tree automata. *Theoretical Computer Science* **22**(1983): 219–232.

[Th1]   A. Thue. Über unendliche Zeichenreihen. *Skrifter utgit av Videnskapsselskapet i Kristiania*. I (1906): 1–22.

[Th2]   A. Thue. Probleme über Veränderungen von Zeichenreihen nach gegebenen Regeln. *Skrifter utgit av Videnskapsselskapet i Kristiania* I **10**(1914).

[Tu]   A. M. Turing. On computable numbers, with an application to the Entscheidungsproblem. *Proceedings of the London Mathematical Society* **42**(1936): 230–265.

[We]   E. Welzl. Color-families are dense. *Theoretical Computer Science* **17**(1982): 29–41.

[Wo]   D. Wood. *Grammar and L Forms: An Introduction*. Berlin: Springer-Verlag (1980).

[Y]   A. C. Yao. Protocols for secure computations. *Proceedings of the 23rd Annual FOCS Symposium* (1982): 160–164.

Index

ENCYCLOPEDIA OF MATHEMATICS AND ITS APPLICATIONS
GIAN-CARLO ROTA, *Editor*

Other volumes in preparation